Postharvest Management and Processing of Fruits and Vegetables — *Instant Notes*

Postharvest Management and Processing of Fruits and Vegetables — *Instant Notes*

by :

SATISH KUMAR SHARMA
Assistant Professor, Horticulture
(Fruit & Vegetable Processing)
College of Forestry & Hill Agriculture
G.B. Pant University of Agriculture & Technology
Hill Campus, Ranichauri, Tehri (Garhwal)
Uttarakhand-249199

New India Publishing Agency
Pitam Pura, New Delhi- 110 088

Published by
Sumit Pal Jain *for*
New India Publishing Agency
101, Vikas Surya Plaza, CU Block, L.S.C. Mkt.,
Pitam Pura, New Delhi- 110 088, (India)
Phone: 011-27341717, Fax: 011-27341616
Mobile : 09717133558
E-mail: newindiapublishingagency@gmail.com
Web: www.bookfactoryindia.com

ISBN : 978-93-80235-20-2

Typeset at: Typographiya # 98 11 48 23 28
Printed at: Jai Bharat Printing Press, Delhi

G.B. Pant University
of Agriculture and Technology,
PANTNAGAR - 263 145
Distt. Udham Singh Nagar
Uttarakhand

Dr B.S. Bisht
Vice Chancellor

Foreword

India is second largest producer of fruits and vegetables in the world due to its rich biodiversity and varied agro-climatic conditions, spread from true tropical to extreme cold temperate region. Unlike other agricultural crops, the horticultural crops especially fruits, vegetables, flowers and ornamentals are highly perishable in nature having limited shelf life. It a paradox that inspite of the fact that India is leading in production of horticultural crops, the post harvest losses in our country are also amongst the highest (20-40%) in the world. In order to reduce these losses, that deprives a large proportion of our population from getting sufficient nutrition primarily in terms of vitamins and minerals, all of us including farmers, researchers, academicians, industry and government have to join hands and make efforts to save what we produce. This continuously necessitates producing well trained manpower having technological know how and acquaintance with the latest developments in the field of post harvest technology.

The book on "Postharvest Management and Processing of Fruits and Vegetables" is a well written text book covering a great deal of ready information on post harvest physiology, processing and value addition of horticultural crops. Besides this, latest technologies i.e. irradiation, pulsed electric fields, magnetic fields, pulsed light treatment, high pressure technology, ultrasonics, nutraceuticals *etc.* have also been briefly introduced in the book.

The book shall prove to be very useful to the post harvest technologists, horticulturalists, food scientists, entrepreneurs, supply-chain management professionals, warehouse handlers, teachers, undergraduate and postgraduate students of various agricultural universities. I compliment the author for making a sincere effort for bringing out this useful publication.

(B.S. BISHT)
Vice Chancellor

Preface

Postharvest Technology is a very important branch of agriculture and its importance further increases in the filed of horticulture as the horticultural crops are highly perishable in nature. Although postharvest technology is being practised in all parts of the world since time immemorial, the industry is still in infancy in India. Fruits, vegetables and flowers are produced in many parts of the country but cold chain is missing. Marketing channels are long and delay movement of commodity from producer to consumer. Processing and preservation is also negeigible under organised sector. Therefore, undertaking continuous research, teaching and training becomes imperative in today's changing era of science. As such, there is no book available which discusses the basics of the subject as instant notes on various topics pertaining to the postharvest management of horticultural crops. The students, teachers and researchers often search for a direct reference which is complete on the subject for teaching the undergraduate students or postgraduate beginners.

The author therefore made an attempt to meet out this need. This book on "Postharvest Management and Processing of Fruits and Vegetables" has been divided into 4 parts. Part I deals with Postharvest Physiology, Part II with Processing, III with Novel Technologies and Part IV with Objective questions. Effort has also been made to add few recent topics i.e. pulsed electric fields, magnetic fields, plused light treatment, high pressure technology, functional foods, designer foods, GM foods, HACCP, nutraceuticals under part III alongwith objectives questions that are often asked in various competitive examinations. Contents have been covered considering the syllabus of basic course on *postharvest management and processing of fruits and vegetables* being taught in most of the agricultural universities in India at undergraduate and postgradute levels.

I hope that the students, researchers, teachers and all those who have interest in the subject would welcome the book and find it very useful. I also thankfully acknowledge the information on various topics obtained from literature and websites (mentioned in the reference section) that were consulted during the preparation of this manuscript.

Author

Common Abbreviations

APEDA	Agricultural and Processed Food Products Export Development Authority
a_w	Water activity
BIS	Bureau of Indian Standard
CA storage	Controlled atmospheric storage
CAC	Codex Alimentarius Commission
CFB Cartons	Corrugated fibre board cartons
EUREPGAP	Euro-Retailer Produce Working Group Good Agricultural Practices
FPO	Fruit Products Order (1955)
GAP	Good agricultural practices
GHP	Good Hygiene Practices
GMP	Good manufacturing practices
Gy	Gray
HACCP	Hazard Analysis Critical Control Point
HTST	High temperature short time
ISO	International Standards Organisation
KMS	Potassium metabisulphite
LTLT	Low temperature long time
MA storage	Modified atmospheric storage

MFPO	Meat and Food Products Order (1973)
MMPO	Meat and Meat Products Order (1992)
°B	degree brix
Pa	Pascal
PFA	Prevention of Food Adulteration Act (1954)
ppm	Parts per million
PPO	Polyphenol oxidase
psi	Pounds per square inch
RTE	Ready to eat
RTS	Ready to serve
SO_2	Sulphur dioxide
SS	Stainless steel
TSS	Total soluble solids
VPO	Vegetables Product Order (1967)
ZECC	Zero energy cool chamber
TBT	Technical Barriers to Trade
WTO	World Trade Organisation
SPS	Sanitary and Phytosanitary
GM foods	Genetically Modified Foods

List of Difference Tables

Contents

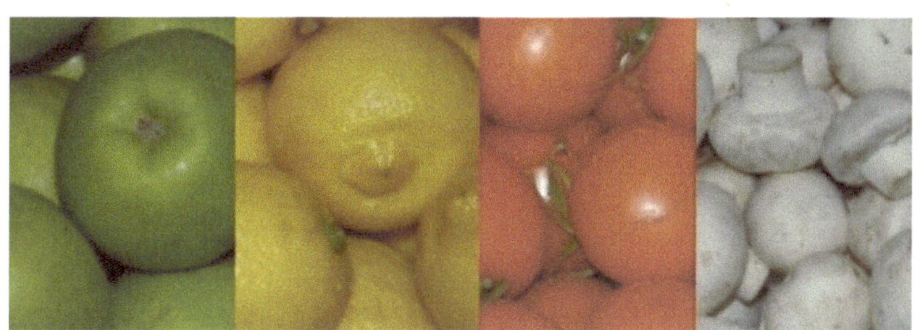

PART I
Postharvest Physiology

1

Importance of Postharvest Technology

India, one of the largest food producer countries (I in milk production, II in fruit and vegetable production and III in food grain production at world level), possesses the potential of being the biggest with the food and agricultural sector contributing around 26 percent of the Gross Domestic Product (GDP). It is a paradox that people of the country producing such huge quantities of food including fruits, vegetables, milk, cereals grains etc. are still suffering from many nutritional deficiencies in their diet, and even in some places, the stomachs of people remain unfilled for the want of food. The credit of this would predominantly go to the poor postharvest infrastructure as well as knowhow and accessibility to the facilities. All this makes us one among those countries having largest postharvest losses. The reduction of losses occuring during food handling, processing, storage, distribution and use would be an indirect way of increasing the food production and making the country self-sufficient in nutritional requirements.

Fruit : Botanically fruits are ripened ovary. Fruit is a plant part that may or may not be purely derived from ovary of the flowers, may be fleshy or pulpy in character, often juicy and predominantly sweet with some what acidic taste, fragrant aromatic flavours and generally not consumed cooked during principal part of our meals.

Classification of Fruits

a. *Simple Fruits*: Fruits which develop from a single ovary of single flower. eg. lemon, lime, citrus, peach, apple, pear etc.
b. *Aggregate Fruits*: Fruits which develop from many ovaries of a single flower. eg. raspberries, strawberries, black berries.
c. *Multiple Fruits*: Fruits which develop from many ovaries of many flowers. eg. pineapple

Vegetable : These are the plant parts which are less sweet, generally non-acidic and consumed either cooked or raw during the principal part of our main meals. Vegetable may be fruit, root, stem, leaf or flower of plant or their modifications.

Difference between Fruit and Vegetable

Fruit	Vegetable
1. Plant part generally formed from flower (flower part) or inflorescence	1. It can be leaf, stem, root, flower etc.
2. Generally consumed as raw, not during principal meals	2. Generally consumed as cooked during principal meals
3. Generally acidic in nature	3. Generally non-acidic type food
4. Mostly woody perennials	4. Mostly non-woody annuals on biennials
5. Mostly propogated asexually	5. Mostly propagated by seed.

Production Statistics of Fruits and Vegetables in India

Fruits total excluding melons

Year	Area (ha)	Area (million ha)	Production (tonnes)	Production (million tonnes)
2007	4398400	4.40	51141800	51.14
2006	4165500	4.17	48045000	48.05
2005	4629075	4.63	42462400	42.46
2000	3800275	3.80	41903120	41.90
1990	2513210	2.51	27358959	27.36
1980	2206514	2.21	20357397	20.36
1970	1822084	1.82	15786680	15.79
1961	1549170	1.55	13372500	13.37

Vegetables and melons total

Year	Area (ha)	Area (million ha)	Production (tonnes)	Production (million tonnes)
2007	5674200	5.67	72544600	72.54
2006	6062400	6.06	75933600	75.93
2005	7913600	7.91	91687500	91.69
2000	5463600	5.46	72283700	72.28
1990	4799957	4.80	48936575	48.93
1980	4350600	4.35	35975100	35.97
1970	3489170	3.49	25985900	25.99
1961	2779350	2.78	18468500	18.47

Source: FAO, 2009

Definition

Postharvest technology/Postharvest management may be defined as the branch of agriculture that deals with all the operations right from harvesting or even the pre-harvest stages till the commodity reaches consumer, either in fresh (grains, apple, mango, tomato fruits) or processed form (flour, juice, nectar, ketchup) and utilization of the wastes (pomace, peel, seed, skin etc.) in a profitable manner (manufacture of fermented beverages, colour extraction, pectin extraction etc.).

Whatever unit operations are done with the crops right from the harvesting stage till the product is consumed, all is dealt with in postharvest technology. Sometimes some specific pre-harvest operations i.e. pre-harvest sprays of calcium and boron on fruits that results in improving firmness (postharvest quality) are also dealt with in postharvest technology as these are not the regular agronomic practices but are done to affect the postharvest quality. In postharvest technology we deal with 3 types of products i.e. fresh produce, processed products and handling and processing wastes.

Importance of Postharvest Technology

Worldwide postharvest fruit and vegetables losses are as high as 25 to 40% and even much higher in some developing countries. Reducing postharvest losses is very important, for ensuring that sufficient food, both in terms of quantity and quality is available to every individual in this world. The world population will grow from 5.7 billion (1995) to 8.3 billion in 2025 so, the requirement of food will also increase. Production of a particular crop can be increased to a limited extent by using high yielding varieties and better agronomic practices. Wasting 40% of the production in a certain crop is just like throwing away Rs 4,000/- out of a total salary of Rs 10,000/-, which nobody would ever like to do. Then why do we accept 40% postharvest losses in horticultural crops? For the last many years we have heard and read in many periodicals and books that India ranks second in production of fruits and vegetables in the world, but if we subtract the quantity lost due to postharvest losses, our rank in production of fruits and vegetables in the world would be much lower than what it is at present. Following points shall highlight the importance of postharvest technology.

a. *Reduction of postharvest losses:* Postharvest technology ensures reduction of losses in what has already been produced. So, reduction of postharvest losses is an alternative way of increasing production of agricultural and horticultural crops.

b. *Reduction of cost of production:* Postharvest technology reduces cost of production, packaging, storage, transportation, marketing and distribution, lowers the price for the consumer and increases the farmer's income.

c. *Reducing malnutrition:* Proper postharvest technology ensures availability of sufficient food to all thus reducing malnutrition and ensuring healthy growth of the nation. It also extends the season of availability of a particular commodity.

d. *Economic loss reduction:* Reduces economic losses at growers level, during marketing and at consumer's end.

e. *Availability:* Had there been no knowledge of postharvest technology, apples would not have ever reached Kerala and banana in Himachal Pradesh or Kashmir. Today we can get perishable commodities like banana, tomato etc., throughout the year and in almost every place in the country. Apples can be made available through out the year although the cropping season is just for 2-3 months. Thanks to the advancements made in the field of postharvest technology. The increasing exports of fruits and vegetables has become possible only by the interventions made in postharvest technology.

f. *Employment generation:* The food processing industry ranks first in terms of employment generation with approximately 15 lakhs persons employed. Employment potential in postharvest and value addition sector is considered to be very high. Every Rs. 1 crore invested in fruits and vegetable processing in the organized sector generates 140 persons per year of employment as compared to just 1050 person day of employment per year in Small Scale Investment (SSI) units. The SSI unit in food industry employs 4,80,000 persons, contributing 13% of all SSI units employed.

g. *Export earnings:* Export of fresh and processed horticultural commodities also attracts valuable foreign exchange.

h. *Defence and astronaut's requirements*: Defence forces posted in remote border areas as well as astronauts who travel into space have special requirements of ready to eat and high energy low volume food. This requirement is fulfilled by processing industries.

i. *Infant and sports preparations*: Today special infant and sports drinks and other processed preparations are available for use especially by these people. These preparations are done especially to meet the specific nutritional requirements of their body.

Postharvest Losses

Fruits	Estimated losses (% of total production)	Vegetable	Estimated losses (% of total production)
Papaya	40-100	Potato	35-100
Citrus	20-95	Sweet potato	35-95
Banana	20-80	Lettuce	62
Avocado	43	Tomato	5-50
Stone fruits	28	Cauliflower	49
Grape	27	Carrot	44
Apple	14	Cabbage	37
Overall	30-40 % in different crops	Onion	16-35

Source : Verma and Joshi, 2000

Possible Causes of Postharvest Losses

A significant portion of the world's food harvest is lost to spoilage and infestations on its journey to the consumer. In developing countries, where adverse weather and poorly developed infrastructure contribute to the problem, losses are sometimes of staggering proportions. Postharvest losses occur at all stages begining with harvesting through handling, storage, processing, marketing till consumption.

Extent of Postharvest Losses would Depend on the Following Factors

- Perishability of the commodity (more losses in perishables)
- Ambient temperature and RH (high temperatures causing more losses)
- Fungal and bacterial decay (increases at high temperature)
- Damage by pests-insects, rodents and birds
- Time elapsed between harvesting and consumption
- Efficiency of postharvest handling, storage and processing operations.

Reasons for High Postharvest Losses

a. *Unfavourable climate:* India is a country where most of the fruits are available from April to October, which is the hotter part of the year. The high temperature coupled with high humidity during rainy season (June end to Sept) are most congenial for the growth and development of decay microorganisms.

b. *Poor cultural practices*: In India, farmer does not know that who will be the ultimate buyer or consumer of his produce. He is just producing a particular commodity to sell and earn his living. No cultural practices are practiced to improve the postharvest quality and shelf life. All cultural practices done, aim at improving production and control diseases and pests.

c. *Improper storage conditions*: We do not have sufficient cold storage facility to store our perishables during the glut period. The "cold chain" in which the commodity is handled at low temperatures from harvesting till consumption and which is often recommended for handling perishables is hardly followed in India. Rather, we sometimes give temperature shocks to the commodity. The commodity is harvested during day time when it is hot, transported in trucks or by railway (un refrigerated), sold in wholesale markets (in hot), then put into cold storages (some perishables), again transported to retail market and sold there in hot, purchased by consumer and put into refrigerators at home and then consumed.

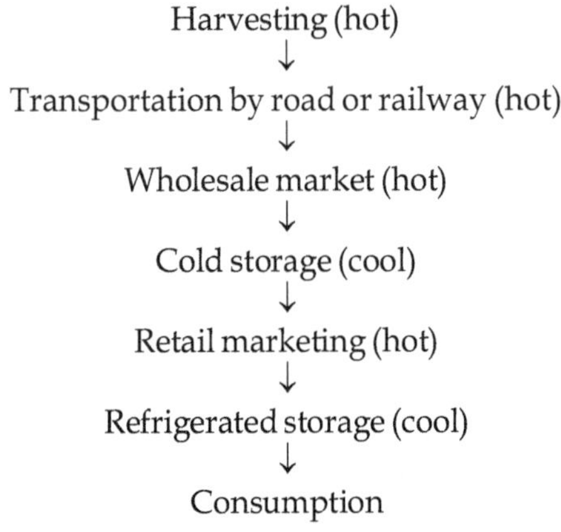

d. *Inadequate handling during transportation:* Harvesting in many of the crops is not proper and handling during packaging and transportation is rough and non-scientific thus causing a lot of bruising and impact damage to the commodity and inviting decay during marketing and storage.

e. *Inadequate facilities available for processing*: India processes less than 2% in organized sector as against 50-70% in some developed countries. The reason is inadequate processing facilities and under-utilization of installed capacity.

f. *Government policies*: This involves the political conditions under which a technological solution for reducing postharvest losses fails to be effective or is unimplementable, eg. lack of a clear policy for facilitating and encouraging utilization of fruits, vegetables and other horticultural perishables and administration of human, economic, technical and scientific resources to prevent postharvest losses.

g. *Education and resources*: The human, economic and technical resources and efforts for developing programs for prevention and reduction of postharvest losses of perishables are inadequate. There is a lack of awareness and knowledge of technical and scientific technologies associated with preservation, processing, packaging, transporting, marketing and distribution of horticultural commodities.

h. *Transportation facilities* : Vehicles used in transporting raw fruits and vegetables in bulk, to the distant markets are not equipped with refrigeration systems. Raw fruits and vegetables when exposed to high (fluctuating) temperatures and humidity during transportation experience moisture loss, softening and bruising of tissues, inviting pathogens and ending in rotting.

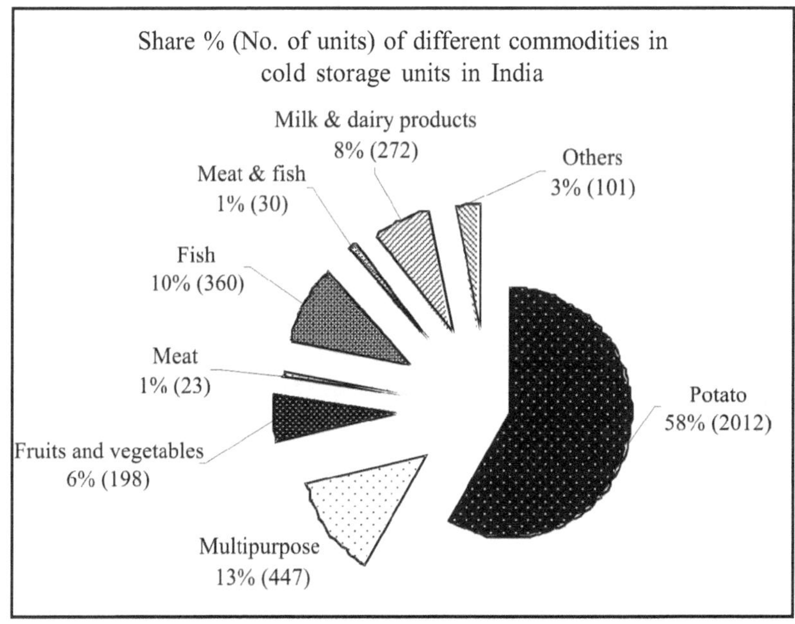

Cold Storage Capacity in India

Source: Anonymous, 2008a

Type of Postharvest Losses

a. Loss of moisture, wilting, shriveling etc.
b. Loss of carbohydrates, vitamins, proteins etc.
c. Physical damage through pest, disease, insect/rodents (rotting etc.)
d. Quality loss due to physiological disorders (chilling injury, freezing injury etc.)
e. Fibre development (beans)
f. Greening (potatoes), sweetness loss (pea)
g. Sprouting (root, shoot growth)
h. Seed germination
i. Development of off flavours
j. Changes in surface colour (green vegetables).

Functions

Postharvest technology aims to-

1. Provide longer shelf life to perishable produce
2. Add value in terms of nutrition and money
3. Maintain and /or improve quality
4. Enhance form, space and time utility of the produce for food, feed, fibre, fuel and industrial purposes.
5. Add variety to the regular food
6. Add convenience for utilization
7. Generate employment and nutritional security
8. Reduce losses and indirectly increase total food production
9. Build a healthy nation.

Postharvest Operations Include

- Harvesting,
- On-farm handling,
- Cleaning, timming, peeling

- Grading, sorting
- Moisture conditioning,
- Pre-cooling
- Waxing,
- Milling,
- Extraction, pulping
- Heating (pasteurization, sterilization, blanching),
- Cooling (freezing),
- Roasting,
- Puffing,
- Flaking,
- Retort processing
- Packaging, wrapping
- Transport and storage

Processing : The fundamental principle of preservation of food by application of heat is known as processing or more precisely "Heat processing". *Processing* is any conscious unit operation resulting in minor or major alterations in the natural shape, size, form, colour, sensory and chemical characteristics of a natural produce and that adds value in terms of preserving or improving the quality and presentation of the produce and convenience of its use.

Preservation : Extending shelf life of fresh or processed food by using various methods of preservation i.e. asepsis, low temperature, high temperature, drying, chemicals, irradiation, high pressures, electric fields, fermentation, anaerobic conditions, packing etc.

FAO has Categorized Food Processing into 3 Sectors

a. *Primary Processing*: involving basic processing of natural produce i.e. cleaning, grading, sorting, washing, dehusking *etc.*

b. *Secondary Processing*: involving elementary modifications of natural foods i.e. packing, waxing, pulping, juice extraction, preparation and preservation of semi-finished products for later use, hydrogenation of edible oils etc.

c. *Tertiary Processing*: involving high levels of modifications to considerably alter the natural produce and to make it ready to eat. i.e. ketchup, RTS beverages, ice creams etc.

As far as reduction of postharvest losses is concerned, adoption of primary and secondary processing can alone create a marked difference and would open a number of new avenues for the tertiary processing to be followed.

Today the food processing sector in India has been facing the problems of erratic, inadequate and fluctuating supply of good quality raw materials, inadequate infrastructure, inadequate investment in organized sector, fragmented and unorganized research and development, inadequately trained human resource, lack of quality testing and certification laboratories, lack of short marketing channels, high costs of working capital and taxation and under-utilization of the installed capacity.

Status of Food Processing Industry in India

Food processing is a large sector that covers activities such as agriculture, horticulture, plantation crops, animal husbandry and fisheries. It also includes other industries that use agricultural inputs for manufacturing of edible products. Food processing industry which is worth Rs 350 thousand crores including Rs. 99 thousand crores worth of value added products ranks 5th in terms of production, consumption, export and expected growth in the country.

Processed food industry in India contributes 6.3% of the GDP, and accounts for 13% of export and 6% of the capital investment. India produces about 600 million tonnes of farm produce annually as detailed in the Table ahead.

Production Statistics of Different Agricultural Commodities

Commodity	(in million tonnes)	Rank in the world
Food grains	210	3rd
Oilseeds	26.1	
Sugarcane	232.3	
Fruits	51.1	2nd
Vegetables	72.5	2nd
Milk	90.7	1st
Fish	6.3	7th
Eggs	45,000 million Nos.	5th
Poultry	489 million stock	6th
Livestock	483 million Nos.	1st
Meat	6.0	
Honey	9,000 t	
Mushrooms	40,000 t	
Spices	4.1	
Tea	0.83	
Coffee	0.281	
Loose flowers	0.73	
Cut flowers	2060 million Nos.	

Food Processing Industry (FPI) has been recognized as a sunrise industry of the country. Estimated to be about USD 70 billion in 2007, the industry is heading for a fast growth. The important sub sectors of the FPI are fruits and vegetables, milk and milk products, grains, packaged drinks, beer and alcoholic beverages, meat and poultry, fisheries and packaged & convenience foods. Considering the meager level of processing being done at present, the FPI of the country possess a lot of opportunity and scope for growth. It also provides an opportunity to the growers and other stake holders in the agricultural sector to enhance their income through postharvest processing value addition and export. Among all the individual processed food items, milk products account for maximum value at Rs 4 billion in 2006-07. Fruits and vegetables together constituted Rs.25 billion in 2007. However, export of processed feed like processed meat, milk products, animal casings do fluctuate and do not rise constantly.

Export of Different Processed Food Items from India (Rs billion)

Item	2005-06	2006-07
Fruits and vegetables	24.5	25
Processed meat	0.024	0.068
Milk products	5.5	4
Other processed Food	26.3	36.3

Source : APEDA.

Government Initiatives

- The national policy on food processing aims to increase the level of food processing in organized sector from 2% to 10% in 2010 and 25% in 2025.
- For 100% export oriented units government has decided to levy zero import duty on capital goods.
- For new industries in fruit and vegetable processing, income tax rebate of 100% on profits made during initial 5 years and 25% rebate in the following 5 years has been decided.
- Government also proposes to build terminal markets in 8 major cities i.e. Mumbai, Nasik, Nagpur, Chandigarh, Rai, Patna, Bhopal and Kolkata
- 60 agri-export zones are to be setup for end to end development of exports
- 53 food parks are approved for enabling small and medium food and beverage units.
- National Bank for Agriculture and Rural Development (NABARD) has set aside Rs. 1,000 crore, especially for agro-processing infrastructure and market development.

- Besides reducing excise duty upto zero in some products, cess on coffee export @ Rs. 500 per tonne was abolished. Additional excise duty levied on tea during 2005-2006 is also withdrawn
- Presently, different sectors of food and food processing are regulated by 13 different laws and regulatory orders. Efforts are being made to integrate them.

Level of Processing in Food Processing Sector in India

Item	Level of processing in organized sector	Level of processing in unorganized sector	Total processing
Fruits & Vegetables	1.2%	0.6%	1.8-2.2 %
Milk & Milk products	15%	22%	35-37 %
Meat	21%		21 %
Poultry	6%		6 %
Marine fisheries/ products	1.7%	9%	8 - 10.7%
Shrimps	0.4%	1%	1.4%

Source: Cygnus Report, Indian Food Processing Sector, 2005, MoFPI 2005

Public Private Investment in Postharvest Management and Agricultural Marketing in India During 2006 (in Rs billion)

	Public sector	Private sector	Total Investment
Agri export zone	2	4	6
Cleaning and grading	19	1	20
Cold storage	68	202	270
Fruit and vegetable market	10	-	10
Processing and value addition	375	1125	1500
Refer van	1	5	6
Rural periodic market	21	-	21
Storage	27	27	54

Source: Anonymous, 2008b

Level of Processing in Fruits and Vegetables in Different Countries

UK and USA	60%
Malaysia	80%
Thailand	30%
India	1.8-2.2 %

There are 7,521 regulated markets and 27,294 rural periodic markets (Haat) in the country for handling agricultural produce.

The estimated installed capacity of fruit and vegetable processing industries has increased from 20.8 lakh tonnes (1998) to 21.0 lakh tonnes (1999). The production of processed fruits and vegetables in the country has increased from 9.4 lakh tonnes in 1998 to 9.8 lakh tonnes in 1999 during which the number of licenses issued under Fruit Product Order (FPO), 1955 has increased to 5198 from 5112.

Important Processed Products Manufactured in India

Fruit pulps and juices, fruit based ready-to-serve beverages, canned fruits and vegetables, jams, squashes, pickles, chutneys and dehydrated vegetables. Recently, products like frozen pulps and vegetables, frozen dried fruits and vegetables, fruit juice concentrates and vegetable curries in retortable pouches, canned mushroom and mushroom products have been taken up for manufacture by the industry.

Points to Remember

- √ India is the second largest producer of fruits and vegetables and third largest producer of food grains in the world
- √ India is the largest producer of milk in the world
- √ India ranks 6th in poultry production in the world
- √ The processing level of fruits and vegetables in India is 1.8-2.2 %

- √ The national policy on food processing aims to increase the level of food processing in organized sector from 2% to 10% in 2010 and 25% in 2025
- √ Botanically fruits are ripened ovary. *Fruit* is a plant part that may or may not be purely derived from ovary of the flowers, may be fleshy or pulpy in character, often juicy and predominantly sweet with somewhat acidic taste and fragrant aromatic flavours and generally not consumed cooked during principal part of our meals.
- √ Simple Fruits develop from a single ovary of single flower eg citrus, peach, apple, pear
- √ Aggregate Fruits develop from many ovaries of a single flower eg. raspberries, strawberries
- √ Multiple Fruits develop from many ovaries of many flowers. eg. pineapple
- √ Vegetables are the plant parts which are less sweet, generally non-acidic and consumed either cooked or raw during the principal part of our main meals. Vegetable may be fruit, root, stem, leaf or flower of plant or their modifications.
- √ Postharvest technology may be defined as the branch of agriculture that deals with all the operations right from the harvesting or even the pre-harvest stages till the commodity reaches the consumer either in fresh or processed form and utilization of the wastes in a profitable manner.
- √ NABARD is National Bank for Agriculture & Rural Development
- √ Processing is any conscious unit operation resulting in minor or major alterations in the natural shape, size, form, colour, sensory and chemical characteristics of a natural produce and that adds value in terms of preserving or improving the quality of the produce and convenience of its use.
- √ Preservation is extending shelf life of fresh or processed food by using various methods of preservation i.e. asepsis, low temperature, high temperature, drying, chemicals, irradiation, fermentation, high pressures, electric fields, anaerobic conditions, packing etc.

- √ FAO has categorized food processing into 3 sectors i.e. primary processing, secondary processing and tertiary processing
- √ Food Processing Industry (FPI) has been recognized as a sunrise industry of the country
- √ Highest number of cold storages (2012) out of the total in the country are being used for potato in India.

2

Maturity and Ripening

Physiological Development

Fruits and vegetables can be divided into 3 stages based on their physiological development i.e. growth, maturation and senescence.

Growth: This generally refers to cell division, enlargement and differentiation ultimately giving a particular size, weight and volume to the commodity.

Maturation: It is initiated before the ceasation of growth and lasts till the onset of senescence.

Senescence: The stage when anabolic (synthetic) processes almost terminate and catabolic (degradative) processes are initiated and speeded up causing ageing and finally death of tissues.

Ripening: Stage that begins with the last stages of maturation and lasts till the beginning (first stage) of senescence. Ripening is a very important event in the life of a plant as it transforms a physiologically mature but inedible (or not preferred to be eaten) plant part into an edible, visually and olfactorily attractive and tasty commodity. Ripening marks the completion of developmental phase and is an irrevessible event.

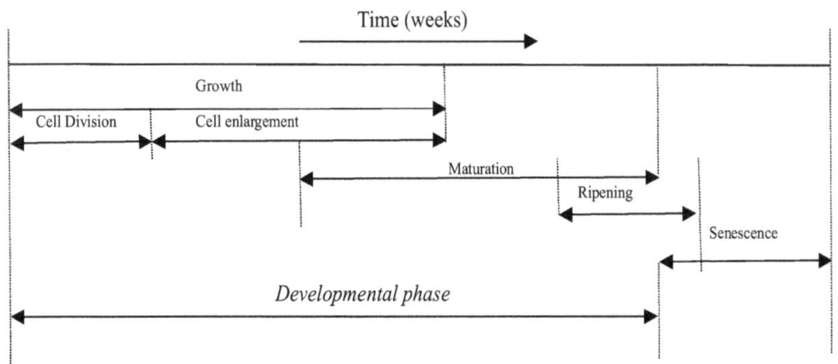

Fig: Physiological Development Stages in Fruits and Vegetables

Generally the growth and maturation of the fruit are completed on the plant itself but ripening and senescence may take place even after detachment of the commodity from the plant. i.e ripening and senescence can take place both on and off the plant.

Important Physico-Chemical Changes Occurring During Ripening

The following changes may take place during the ripening of fruits

- *Maturation* of seed/change in seed colour to brownish (in most of the crops)
- *Changes in fruit skin colour / development of carotenoids*: Generally from green to orange/yellow colour etc. depending on the type of fruit. Desirable in fruits i.e. apricots, peaches, citrus. Red colour development in tomato is due to lycopene (carotenoid)

- *Loss of chlorophyll*: Desirable in fruits but not in most of the vegetables
- *Development of anthocynins*: Red and blue water soluble colour pigments. Desirable in fruits. i.e. apple, cherry, strawberry and plum.
- *Formation of abscission layer* i.e. the mark from where the fruit shall detach from the parent plant.
- *Changes in respiration rate:* Usually respiration rate decreases with the advancement of ripening in non-climacteric fruits, while in climacteric fruits, it rises initially and then declines.
- *Changes in ethylene production*: Usually ethylene production rates increase with the advancement in ripening
- *Softening*: Generally the fruits tend to soften with the advancement in ripening due to changes in the composition of cell wall pectic substances
- *Changes in carbohydrates*: Starch gets converted into sugars.
- *Organic acids* are converted into other substances
- *Flavour changes*: Different aroma volatiles are also produced during the process of ripening
- In some fruits like apple, plum, wax is developed on the surface of fruit

Respiration : The oxidative breakdown of more complex material i.e. carbohydrates (starch, sugars, organic acids), proteins, fats, into simpler molecules, such as CO_2 and H_2O, with the concurrent production of energy and other molecules which can be used by the cells for synthetic reactions.

$$C_6H_{12}O_6 + 6CO_2 = 6CO_2 + 6H_2O + \text{energy}$$

2 types (i) aerobic respiration (occurs in the presence of O_2)
(ii) anaerobic respiration (occurs in the absence of O_2)

Respiration is reverse to *photosynthesis* in which energy derived from Sun is stored as chemical energy in the green plants.

Vital Heat: During the process of respiration, food reserves are converted into energy. Some energy produced during this process is utilized for maintaining life processes and excess of energy is released in the form of heat is called "Vital Heat". This heat should also be considered during any temperature management activity.

Respiration rate: Rate of consumption of oxygen or evolution of CO_2 from a commodity is called as respiration rate. It is an indicator of the metabolic activity of the fruit tissues thus is useful in the determination of potential storage life of produce. Immature fruits and vegetables have highest respiration rates which decline steadily towards maturation, ripening and senescence. It can be measured in ml CO_2 $Kg^{-1}h^{-1}$.

Classification of Fruits, Vegetables and Flowers on the Basis of Respiration Rates

Category	Respiration rate	Crops (at 5°C) ($mgCO_2kg^{-1}h^{-1}$)	Crops (at 15°C) (ml $CO_2kg^{-1}h^{-1}$)
Very low	Less than 5	Dried fruits, nuts, dates	May be nuts
Low	5-10	Apple, citrus, grape, kiwifruit, garlic, onion, potato (mature)	Potato
Moderate	10-20	Apricot, banana, cherry, peach, plum, nectarine, fig, cabbage, carrot, lettuce, pepper, tomato, potato (immature)	Grape, lemon, orange
High	20-40	Avocado, cauliflower, limabean, black berry, raspberry, strawberry	Apple, cabbage,
Very High	40-60	Artichoke, snapbeans, green onion, brussels sprouts, cut flowers	Banana (green), peach, carrot,
Extremely high	More than 60	Asparagus, broccoli, mushroom, pea, spinach, sweet corn	Banana (ripe), lettuce, pea, bean, pear, strawberry

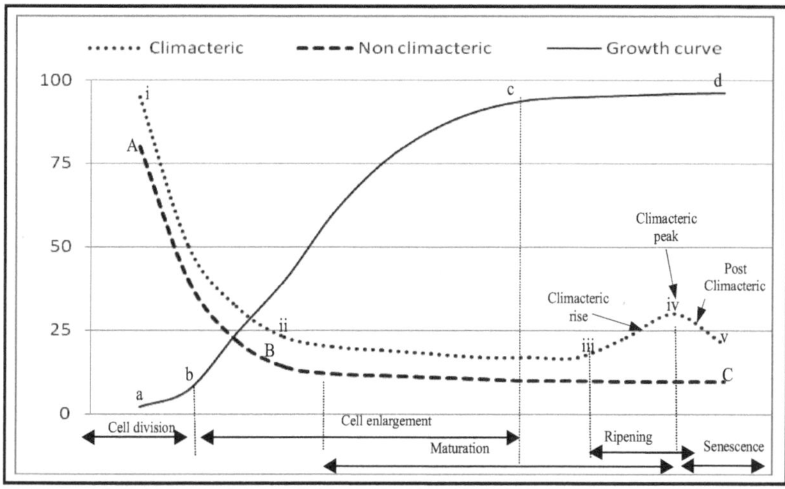

Respiratory Patterns

Source : Wills *et al.* 1996.

Growth Curve

a-b: Cell division, very small size of fruits on the tree

b-c: Cell enlargement, increase in size parameters i.e. length, breadth, diameter, weight, volume etc. of the fruit, maturation starts

c-d: Negligible weight gain in fruit, fruit matures and enters into ripening and senescence at the end.

Climacteric Respiration Curve

i-ii: Sharp decline in respiration

ii-iii Respiration rate becomes steady in maturing fruits.

iii-iv Respiration rate again rises (climacteric rise) and reaches maximum (climacteric maximum or peak). It is very difficult to control the respiration rate in this stage of fruit growth. Storability reduces very fast due to fast ripening and early entry into senescence.

iv-v Respiration rate again drops down (post-climacteric). Fruits enters into senescence.

Non-Climacteric Respiration Curve

A-B: Sharp decline in respiration

B-C: Respiration rate becomes steady in fruits. There is no climacteric rise. Fruits are required to be matured on the tree.

Respiration Quotient (RQ) : It may be defined as the ratio of carbon dioxide produced to that of oxygen consumed during respiration i.e.

$$RQ = \frac{CO_2 \text{ produced (in ml)}}{O_2 \text{ consumed (in ml)}}$$

Complete oxidation of glucose	RQ = 1.0
Complete oxidation of malate	RQ = 1.3
Complete oxidation of stearic acid	RQ = 0.7 (fatty acids have less oxygen per carbon atom in their structure than sugars so require more consumption of O_2 for production of CO_2, thus lower values of RQ)

Therefore, RQ gives an idea about the type of substrate being respired. Low RQ would indicate metabolism of fats and high RQ would indicate metabolism of organic acids i.e more oxygen per carbon atom.

Climacteric Fruits : A sharp increase in respiration (respiratory peak) is shown by the increase in production of CO_2 and decrease in internal O_2 concentration. Ripening in climacteric fruits coincides with the rise in respiratory activity. Coincident with ripening, the climacteric fruits produce much larger amounts of ethylene than non-climacteric fruits. The respiratory climacteric as well as the complete ripening process, may proceed while the fruit is either attached to or detached from the plant.

eg. Apple, Apricot, Avocado, Banana, Blue berry, Breadfruit, Cherimoya, Feijoa, Fig, Guava, Jackfruit, Kiwifruit,

Mango, Muskmelon, Nectarine, Papaya, Passion fruit, Peach, Pear, Persimmon, Plantain, Plum, Sapota, Tomato, Watermelon etc.

Climacteric fruits are those in which the respiration rate is minimum at maturity and remains constant even after harvest which gradually increase at the beginning of ripening followed by sharp rise to a peak (climacteric peak) and then slowly decline (post climacteric stage).

Non-Climacteric Fruits : The fruits that do not exhibit respiratory climacteric (pronounced increase in respiration coincident with ripening forming a peak) are known as non climacteric fruits. Coincident with ripening, the non-climacteric fruits produce much lesser amounts of ethylene than climacteric fruits.

eg. Ber (Jujube), Blackberry, Cashew apple, Cherry, Cucumber, Eggplant, Grape, Grape fruit, Lemon, Lime, Litchi, Loquat, Olive, Orange, Pepper, Pineapple, Pomegranate, Citrus, Satsuma mandarin, Strawberry, Summer squash, Sweet orange, Tree tomato, Tangerine etc.

Non-Climacteric fruits are those that show a gradual decline in respiration rate with ripening. These fruits ripen on the tree and therefore may be harvested when they become edible.

Difference Between Climacteric and Non-Climacteric Fruits

Climacteric	Non-climacteric
1. Fruits exhibit respiratory climacteric (pronounced increase in respiration coincident with ripening forming a peak)	1. Fruits do not exhibit respiratory climacteric
2. Coincident with ripening, the fruits produce much larger amounts of ethylene	2. No such relationship exists in these fruits
3. Fruits may ripen on and off the trees	3. Fruits ripen on the tree only
4. eg. Apple, guava, papaya, tomato	4. eg. Citrus, cucumber, grape, litchi

Factors Affecting Respiration Rate

1. Temperature

During respiration a number of enzymatic reactions take place inside the fruit and the rate of these reactions depend on temperature and may be described by temperature quotient (Q_{10}). Van't Hoff, a Dutch chemist described that the rate of a chemical reaction almost doubles for every 10°C rise in temperature.

Van't Hoff temperature quotient, $Q_{10} = (r_2/r_1)^{10(t_2-t_1)}$

where, r_2 and r_1 are the respective rates of reactions at temperatures t_2 and t_1 in °C. Q_{10} values are the highest at 1-10 °C temperatures and can be as high as 7 but at higher temperatures Q_{10} values are between 2-3.

Example

Suppose the respiration rate of apples is studied at 0°C, 10 °C and 20 °C. At 0°C the respiration rate is r mlCO_2/Kg/hr then according to Van't Hoff temperature quotient

Reaction rate at 0°C = r

Reaction rate at 10°C = 7r

Reaction rate at 20°C = 14r–21r.

In non-climacteric fruits, reduction in temperatures only lowers the rate of deterioration whereas in climacteric fruits, low temperatures can delay the onset of ripening. But in commodities sensitive to *chilling injury* reduction of temperatures below 10° C or more precisely 7 °C for long durations may damage the produce.

There is an inverse relationship between storage life and respiration rate. Higher the respiration rates lesser will be the storage life and *vice versa*. However, there is a direct relationship between temperature and respiration rates upto a temperature range of 30-40° C depending upon the commodity, i.e. higher the temperatures, higher will be the respiration rates and lesser will be the storage life.

2. *Atmospheric Composition*

The concentration of oxygen in the atmosphere surrounding the commodity is very important factor governing the respiration rates. As depicted in the equation earlier in this chapter, respiration involves the breakdown of sugar into carbon dioxide and water in the presence of oxygen. The modern storage structures i.e. controlled atmosphere, modified atmosphere and hypobaric storage are also dependent on the composition of gas and pressure in the surrounding atmosphere of the commodity.

Ethylene is one gas which is produced during the ripening of fruits and is released into the storage atmosphere from the ripening fruits. Presence of ethylene also initiates unwanted ripening in unripe fruits.

3. *Effect of Ethylene*

All fruits produce small quantities of ethylene during ripening but climacteric fruits produce much larger amounts of ethylene than non-climacteric fruits. Internal ethylene concentration of climacteric fruits varies largely but that in non-climacteric fruits experience little change during development and ripening.

Effect of Applied Ethylene on Climacteric and Non-Climacteric Fruits

Climacteric fruits	Non climacteric fruits
- Ethylene applied @ 0.1–1 μL/ Litre for one day is sufficient to hasten full ripening in climacteric fruits	- Applied ethylene @ 0.1–1 μL/ Litre for one day merely increases the respiration and does not hasten full ripening.
- The magnitude of climacteric rise is independent of applied concentrationi.e. 0.1 as well as 1 or 10 μL/ Litre for one day would result in almost equally good response.	- The magnitude of climacteric rise would depend on the applied concentration. 10 or 1000 μL/ Litre for one day would result in more sharp rise as compared to 0.1 or 1 μL/ Litre for one day.
- Rise in respiration in response to applied ethylene occurs only once	- Rise in respiration in response to applied ethylene may occur more than once

Ethylene Biosynthesis

Methionine

↓

SAM (S - adenosyl methionine)

↓ ACC synthase enzyme

ACC (1 amino cyclopropane 1 carboxylic acid)

↓ EFE (ethylene forming enzyme)

Ethylene

Classification of Fruits, Vegetables and Flowers on the Basis of Ethylene Production Rates

Category	Ethylene evolution rate (μL ethylene $kg^{-1} h^{-1}$)	Crops (at 20°C)
Very low	less than 0.1	Artichoke, asparagus, cauliflower, cherry, citrus, grape, ber, strawberry, pomegranate, leafy and root vegetables, cut flowers
Low	0.1 - 1	Blueberry, cranberry, cucumber, brinjal, okra, olive, pepper, persimmon, pineapple, pumpkin, lemon, lime, orange, pineapple, raspberry, watermelon
Moderate	1-10	Banana, fig, guava, mango, tomato, plantain
High	10-100	Apple, apricot, avocado, cantaloupe, feijoa, kiwifruit, nectarine, papaya, peach, pear, plum
Very High	more than 100	Apple, nectarine, cherimoya, passion fruit, sapota

Source : Kader *et al.*, 1985

Biochemistry of Respiration

Different carbohydrates get converted into glucose or other intermediate products of glycolysis.

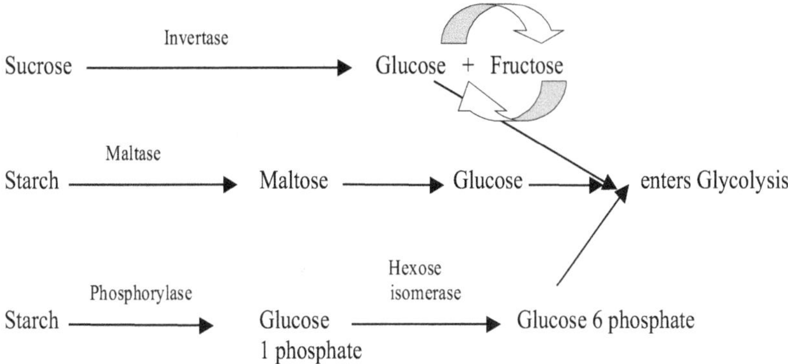

a. Aerobic Respiration

$$C_6H_{12}O_6 + 6CO_2 = \quad 6CO_2 + 6H_2O + \text{energy}$$

The above overall reaction of respiration involving conversion of glucose into carbon dioxide occurs in three main steps

i. *Conversion of glucose to pyruvate by EMP (Embden-Meyerhof-Parnas) pathway / glycolysis*

   glucose + 2 ADP + 2Pi + 2NAD → 2 pyruvate + 2 ATP + 2NADH + $2H_2O$

   Energy released → 2 ATP + 2 NADH = 2 ATP + 2 x 3 ATP = 8 ATP

ii. *Conversion of pyruvate to Acetyl CoA*

   2 Pyruvate + 2 NAD + 2CoA → 2 Acetyl CoA + CO_2 + 2 NADH

   Energy released = 2 NADH = 2 x 3 ATP = 6 ATP

iii. *Conversion of acetyl CoA to CO_2 by TCA cycle:* TCA (tricarboxylic acid) cycle is also called as citric acid cycle or krebs cycle. TCA cycle operates in mitochordria of eukaryotes and cytosol of prokaryotes.

   2 Acetyl CoA + 6 NAD + 2 FAD + 2 GDP → 4 CO_2 + 6 NADH + 2 FADH + 2 GTP

   Energy released = 6 NADH + 2 FADH + 2 GTP
   = 6 x 3 ATP + 2 x 2 ATP + 2 x 1 ATP
   = 18 + 4 + 2 = 24 ATP

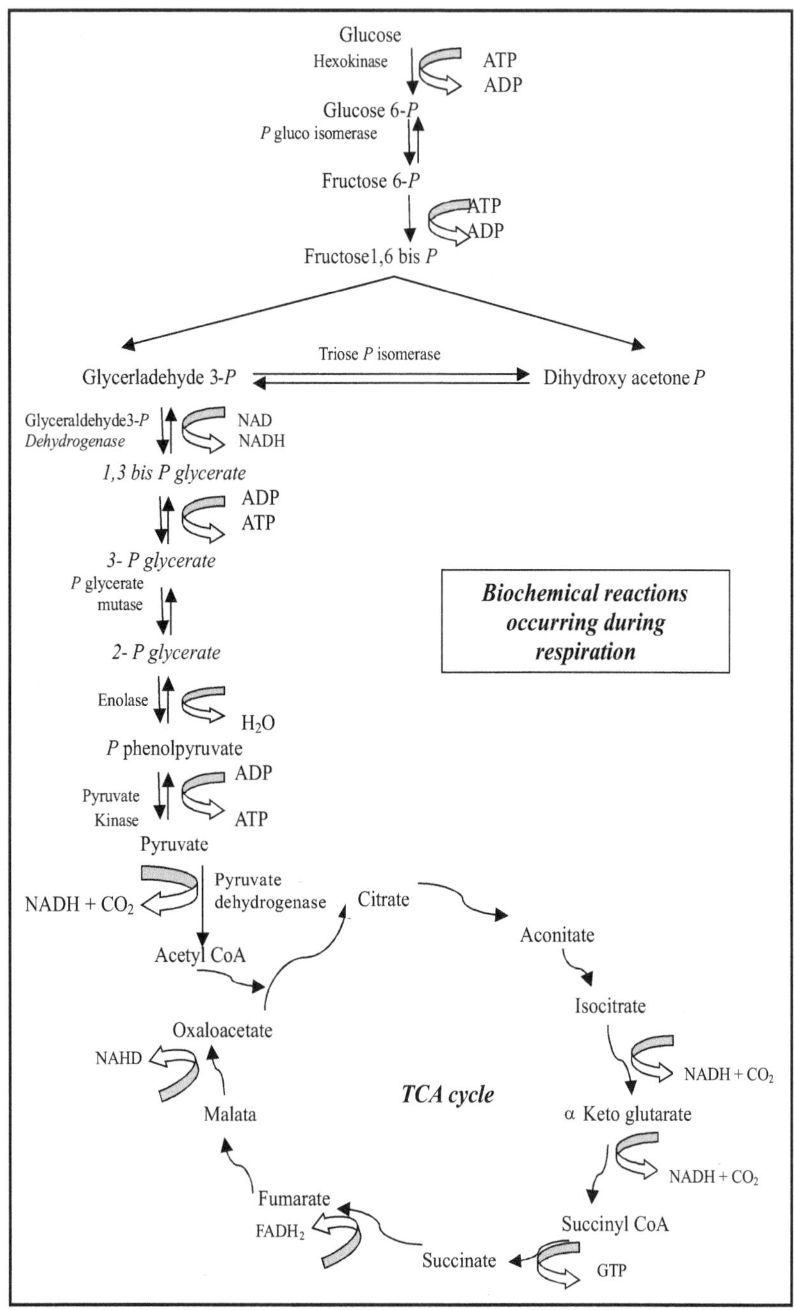

Glucose
Hexokinase
ATP
ADP
Glucose 6-P
P gluco isomerase
Fructose 6-P
ATP
ADP
Fructose1,6 bis P
Triose P isomerase
Glycerladehyde 3-P
Dihydroxy acetone P
Glyceraldehyde3-P Dehydrogenase
NAD
NADH
1,3 bis P glycerate
ADP
ATP
3- P glycerate
P glycerate mutase
2- P glycerate
Enolase
H_2O
P phenolpyruvate
ADP
Pyruvate Kinase
ATP
Pyruvate
Pyruvate dehydrogenase
NADH + CO_2
Acetyl CoA
Citrate
Aconitate
Isocitrate
NADH + CO_2
α Keto glutarate
NADH + CO_2
Succinyl CoA
GTP
Succinate
Fumarate
$FADH_2$
Malata
Oxaloacetate
NAHD
TCA cycle
Biochemical reactions occurring during respiration

Total chemical energy released during oxidation of one molecule of glucose is approximately 1.6 megajoules. About 90% of this energy is preserved in the plant for running the life processes and rest is released as heat.

b. Anaerobic Respiration

Under limited oxygen availability the fruit tissues initiate anaerobic respiration and glucose is converted into pyruvate via EMP pathway and then to lactate or acetalhedyde ultimately forming ethanol. The process is known as fermentation. The concentration of oxygen at which the anaerobic respiration starts is known as ***Extinction Point.***

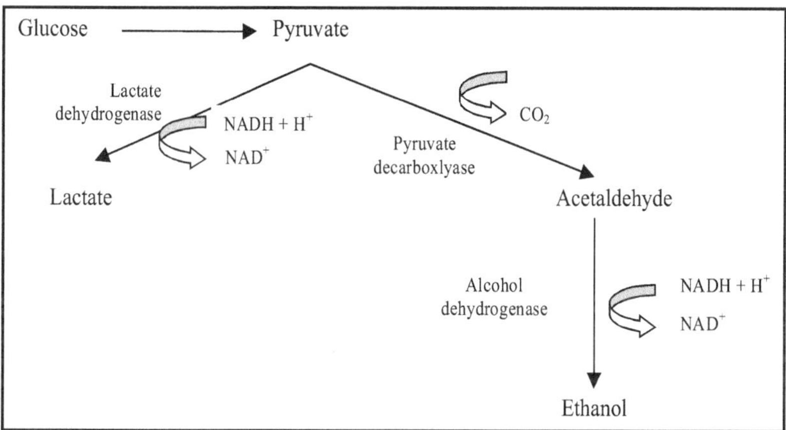

POINTS TO REMEMBER

- √ *Growth* refers to cell division, enlargement and differentiation ultimately giving a particular size, weight and volume to the commodity.
- √ *Maturation* is initiated before the ceasation of growth and lasts till the onset of senescence.
- √ *Senescence* is the stage when anabolic (synthetic) processes almost terminate and catabolic (degradative) processes are initiated and speeded up causing ageing and finally death of tissues.

- √ *Ripening* is the stage that begins with the last stages of maturation and lasts till the beginning of senescence. Ripening marks that completion of developmental phase.
- √ Respiration is the oxidative breakdown of more complex material i.e. carbohydrates (starch, sugars, organic acids), proteins, fats, into simpler molecules, such as CO_2 and H_2O, with the concurrent production of energy.
- √ Respiration is reverse to photosynthesis in which energy derived from Sun is stored as chemical energy in the green plants.
- √ Vital Heat is the energy is released in the form of heat during respiration.
- √ Respiration rate is the rate of consumption of oxygen or evolution of CO_2 from a commodity. Unit = ml CO_2 $Kg^{-1}h^{-1}$
- √ Respiration Quotient (RQ) may be defined as the ratio of carbon dioxide produced to that of oxygen consumed during respiration.
- √ In climacteric fruits there is a sharp increase in respiration (respiratory peak) coincident with ripening and they produce much larger amounts of ethylene than non-climacteric fruits. Ripening may proceed while the fruit is either attached to or detached from the plant eg Apple, Banana, Guava, Kiwifruit, Mango, Papaya, Plum, Sapota, Tomato.
- √ Non-climacteric fruits do not exhibit respiratory climacteric (pronounced increase in respiration coincident with ripening forming a peak) coincident with ripening and produce much lesser amounts of ethylene than climacteric fruits. eg. Ber, Cucumber, Eggplant, Grape, Citrus, Litchi, Loquat, Olive, Pepper, Pineapple, Pomegranate, Strawberry.
- √ Temperature, atmospheric composition and ethylene are the important factors affecting respiration rate of a commodity.
- √ Methionine is the precursor in ethylene biosynthesis.
- √ During aerobic respiration glucose converts into pyruvate via. EMP pathway and then into Acetyl CoA and then brteaks down into CO_2 via TCA cycle.
- √ During anaerobic respiration glucose converts into pyruvate via. EMP pathway and then Lactate or acetaldehyde ultimately producing ethanol. The reaction is known as fermentation.

- The concentration of oxygen at which the anaerobic respiration starts is known as *Extinction Point.*
- Starch and sucrose also converts into glucose or Glucose 6 phosphate before entering into TCA cycle.
- Van't Hoff, a Dutch chemist described that the rate of a chemical reaction almost doubles for every 10^0C rise in temperature and gave a temperature quotient $Q_{10.}$

3

Quality Management for Fresh Marketing

Maturity Standards and Harvesting Indices

Harvesting : Harvesting is detaching a commodity from the point of its origin. This point of origin may be an above ground plant part i.e. shoot eg. apple, tomato etc. or an underground plant part eg potato, carrot, etc. or in water i.e. harvesting of fish from a pond (*not to be covered in this book*).

Goal of Harvesting

- To gather the commodity at optimum stage of maturity depending upon the purpose, with minimum of damage and loss, in as minimum time as possible and at the lowest possible cost.

Methods of Harvesting

1. *Hand harvesting*: Harvesting by hand is being practiced in all the horticultural crops since time immemorial.

Some of the crops eg flowers even today are harvested by hands. But in India hand harvesting is still the most common method used in horticultural commodities due to inadequate mechanization, small land holding and diversity of crops being grown by a small farmer. Some harvesting aids can be used for increasing the efficiency of labour.

Advantages of Hand Harvesting

- Accurate selection for maturity
- Accurate grading (discarding the damaged, diseased fruits at the time of harvesting only)
- Minimum damage to the commodity
- Rate of harvesting can be increased by employing more number persons
- Minimum of capital investment
- Same labour can be used for harvesting different types of crops eg. apple and gladiolus can be harvested by same person but cannot be harvested by same machine.
- Multiple harvests: Immature or small sized fruits could be left on the plant for next harvest eg. pea, capsicum where 3-4 harvests are taken from the same plant.

Disadvantages of Hand Harvesting

- More time consuming
- Dependent on availability of labour. Labour is very costly in some countries like Japan. Strikes of labour during the time of harvesting may result in crop losses.

2. *Mechanical harvesting*: Harvesting by use of machines. It is very useful for rapid harvesting of a particular crop and at low cost. Special harvesting machines are designed for specific crops. In developed countries mechanical harvesting is common for most of the crops, but in India it is still very uncommon.

Advantages of Mechanical Harvesting

- Rapid harvesting, thus saving time
- Less dependency on labour. No risks of labour strikes and labour management related problems.
- Improved working conditions for workers

Disadvantages of Mechanical Harvesting

- Required skilled manpower for use of machine, therefore dependence on trained labour
- Improper machine usage may result in huge economic losses.
- Machine requires regular maintenance.
- Same machine cannot be used for many crops.
- May cause damage to perennial crops (bark branches of trees)
- Social impacts of low labour requirements and employment
- Inadequate facilities for handling and processing large quantity of harvested produce

Precautions to be taken During Harvesting

- There should be minimum damage to the commodity and self
- Method of harvesting should be less expensive
- Less damage to the plant
- Should be done at optimum harvesting stage
- Some fruits i.e. citrus fruits (malta, lemon, orange etc) and temperate stone fruits (plum etc) are reported to have longer shelf life and lesser rotting during storage when harvested alongwith attached pedicel. So such fruits should carefully be harvested with attached pedicel.

Maturity : It is a stage of fruit growth preceding ripening. Physiologically the meanings of "mature" and "ripe" are not same although most common people use these two terms as synonyms.

Maturity can also be defined as a stage at which commodity has attained sufficient physiological development such that after harvesting at that stage and adequate postharvest handling (including controlled ripening if required), its quality would be above the minimum acceptable to the ultimate consumer. In fruits the optimum maturity stage may or may not coincide with stage having optimum eating quality. eg. banana is harvested a lot earlier than it attains eating quality while citrus may be harvested at stage when it attains optimum eating quality. The maturity stage may also be influenced by the ultimate purpose of use of a commodity (would be discussed under Horticultural maturity).

Physiological maturity: Stage at which a plant or a plant part continues ontogeny even if detached from its parent plant or the point of origin. *Ontogeny: Complete developmental history of an organism from egg/spore/bud etc to an adult individual.*

Horticultural maturity: It may be defined as the stage at which a plant or a plant part possesses all the pre-requisites for utilization by the ultimate consumer for a particular purpose. eg. horticultural maturity stage of tomato if harvested for long distance transportation would be the "turning stage of skin from green to red" while the optimum stage of harvesting of the same crop for home use or local markets would be "when the fruits have attained full red colour".

Maturity Indices

Maturity indices are the factors on the basis of which it is decided whether the commodity is ready for harvesting based on its horticultural maturity or not? Generally a single maturity index is not considered to be reliable. In most of the crops more than one or two indices should be made use of while determining the exact stage of optimum maturity. Experience of growing, harvesting and marketing a particular crop alongwith critical observations would be the best for determination of the optimum maturity. A beginner should seek the help of a local experienced person in determination of maturity of a crop till he himself gains experience and confidence.

Types of Maturity Indices

1. Computational Methods

Method	Crops
Calendar date	Most fruits
Days from full bloom to harvest	Apple, pear
Mean heat units	Pea, apple, sweet corn

2. Physical Methods

Method	Crops
Fruit retention strength	Apple
Fruit size & surface morphology:	
• Netting	Melons
• Cuticle formation	Grape, tomato
• Development of wax	Plum, apple
Size, weight etc.	All fruits and vegetables
Specific gravity	Cherries, watermelons, potatoes
Colour (skin, flesh and seed)	All fruits and most of the vegetables
Flesh firmness	Apple, pear, temperate stone fruits
T Stage	Apple
Tenderness	Pea, bean (green)
Juice content	Citrus, orange, lime, lemon etc.
Oil content	Avocado
Development of abscission layer	Melons, apple, feijoas
Solidity	Lettuce, cabbage, brussels sprouts

contd...

Shape:	
- Angularity of fingers	Banana
- Fullness of cheeks	Mango
- Compactness	Cauliflower, broccoli
Degreening of foliage	Ginger
Internal colour and structure	Flesh colour in some fruits (guava)
	Formation of jelly-like mass in tomato

3. Chemical Methods

Method	Crops
TSS	Apple, plum, apricot, pear etc
Titratable acidity & TSS : acid ratio	Pomegranate, citrus, papaya, melons, grapes
Sugars & Sugar : acid ratio	Apple, pear, stone fruits, grapes
Starch	Apple, pear
Astringency / tannins	Persimmon, dates, walnuts, pecan nuts

4. Physiological Methods

Method	Crops
Respiration Rate	Can be used in many fruits and vegetables
Ethylene Evolution Rate	Apple, pears

Postharvest Disorders

These are caused by improper climatic conditions during handling or nutritional deficiencies or imbalance. When a freshly harvested horticultural commodity is exposed to undesirable temperatures it may result in postharvest physiological disorders. There are various types of physiological disorders as discussed below.

a. *Freezing injury* : When a commodity is held below its freezing temperature, this injury occurs. Generally this temperature is 0° C or even less in some fruits.

   The *common symptoms* of this injury are formation of ice crystals inside the fruit, browning of damaged areas.

b. *Chilling injury*: This type of injury occurs in subtropical or tropical fruit crops. Chilling injury occurs when a commodity is held above its freezing point but below its optimum storage temperature (5-15° C). eg. peel of banana when kept in refrigerator turns black after some time.

   The *common symptoms* of this injury are surface and internal discolouration, pitting, water soaked areas, uneven ripening or failure to ripen, undesirable off flavor development, high incidence of surface moulds and decay.

Approx. Lowest Safe Storage Temperatures to Avoid Chilling Injury

Fruit/Vegetable	Temperature (° C)
Banana	12
Cucumber, egg plant, lime, papaya	7
Avocado, mango	5 - 12
Lemon	10
Melon	7 - 10
Pineapple	6 - 12
Tomato	10 - 12

c. *Heat injury*: This injury occurs when the commodity is exposed to direct sunlight or excessively high temperature.

   The *common symptoms* of this injury are bleaching, surface burning or scalding, uneven ripening, softening, desiccation and weight loss.

d. *Bitter pit*: It is common in apples and occurs due to the deficiency of calcium in the fruits.

The *common symptoms* of this disorder are bitterness of the fruit from inside. Fruit looks normal from outside. Pre-harvest sprays of calcium may be helpful in reducing the incidence of this disorder.

e. *Blossom end rot*: It is common in tomato, pepper, watermelon etc. and occurs due to the deficiency of calcium in the fruits.

   The *common symptoms* of this disorder are fruit starts rotting from the blossom end and cause losses. Pre-harvest sprays of calcium may be helpful in reducing the incidence of this disorder.

f. *Superficial scald*: It is common during cold storage of apple and occurs due to oxidation products of alpha farnesene that occur when the natural antioxidants are degraded or inactivated during cold storage.

   The *common symptoms* of this disorder are development of scald (slightly sunken skin discolouration) on fruit surface. Addition of synthetic anti-oxidants effectively prevents this disorder. Postharvest dip in diphenylamine (0.1-0.25%) or 1,2 dihydro, 6 ethoxy, 2,2,4 trimethyl quinoline (ethoxyquin) (0.2 - 0.5%) may be effective in control of the disorder.

Postharvest Diseases

There are a number of microorganisms which cause postharvest damage, decay and rotting in fruits and vegetables. Most of them are weak pathogenst (*Alternaria, Botrytis, Phomopsis, Rhizopus, Penicillium* etc.) which can only invade the bruised or damaged produce and cannot penetrate healthy fruit skin except few i.e. *Colletotrichum* which can even penetrate health fruit skin. It indicates that postharvest decays are most commonly caused only after improper handling of the produce during, harvesting, field handling, transportation and storage.

Important Postharvest Diseases

Fruit /vegetable	Disease	Causal organism
Apple, Pear	Blue mould rot	*Penicillium expansum*
	Grey mould rot	*Botrytis cinerea*
	Lenticel rot	*Gloeosporium album*
Banana	Crown rot	*Colletotrichum musae, Fusarium roseum*
	Anthracnose	*Colletotrichum musae*
Citrus (lemon, orange etc.)	Grey mould rot	*Botrytis cinerea*
	Blue mould rot	*Penicillium italicum*
	Green mould rot	*Penicillium digitatum*
	Stem end rot	*Phomopsis citri, Alternaria citri*
	Sour rot	*Geotrichum candidum*
Grape	Grey mould rot	*Botrytis cinerea*
Peach	Brown rot	*Sclerotinia fructicola or Monilinia fructicola*
	Rhizopus rot	*Rhizopus stolonifer*
Mango	Anthracnose	*Colletotrichum gloeosporiodes*
Cherry	Brown rot	*Sclerotinia fructicola or Monilinia fructicola*
Strawberry	Grey mould rot	*Botrytis cinerea*
	Rhizopus rot	*Rhizopus stolonifer*
Pineapple	Black rot	*Ceratocystis paradoxa*
Potato	Bacterial soft rot	*Erwinia carotovora*
	Dry rot	*Fusarium sp.*
Sweet potato	Black rot	*Ceratocystis fimbriata*
Carrot	Watery soft rot	*Sclerotinia sclerotiorum*
Leafy vegetables	Grey mould rot	*Botrytis cinerea*
	Bacterial soft rot	*Erwinia carotovora*
	Watery soft rot	*Sclerotinia sclerotiorum*

Source : Wills *et al.* 1996

Control of Postharvest Diseases & Pests

1. Reduction in bruising and impact damage during harvesting, handling, transportation and storage
2. *Application of low temperatures*: Low temperatures are very effective in controlling postharvest diseases as it not only

creates unfavourable conditions for the growth of pathogenic microorganismsm but also reduce the respiration rates, thus prolongs the safe storability of the produce. However, the low temperature treatments may not be very suitable for some tropical and subtropical fruits that are sensitive to chilling injury.

3. *Application of heat treatment*: It may be given in chilling sensitive fruits. It may be applied as dry heat (hot air) or hot water dips (at 38-55°C) in *papaya, stone fruits* and *cantaloupes.*

   Advantages of *hot water dip*

   - It is more effective as this not only controls surface infections but also controls microbial growth inside the fruit.
   - It does not leave any chemical residues

4. Irradiation can also be used to control diseases, disorders and pasts

Crop	Control of Disease/disorder/pest	Min. dose (kGy)
Apple	Scald/ brown core	1.5
Apricot, peach, nectarine	Brown rot	2
Banana	Ripening inhibition	0.30-0.35
Lemon	*Penicillium* rot	1.5 -2.0
Mushroom	Inhibition of stem growth and cap opening	2
Orange	*Penicillium* rot	2
Papaya	Disinfestation of fruit fly	0.25
Pear	Ripening inhibition	2.5
Potato	Inhibition of sprouting	0.08-0.15
Strawberry, grape	Grey mould	2
Tomato	Alternaria rot	3

Source : Wills *et al.* 1996

But in most of the above cases, the technology of irradiation finds a limited commercial application as either some cheaper and more effective alternatives are available or the treatment leaves undesirable effect on the produce and cause abnormal ripening.

Relative Tolerance of Fresh Fruits and Vegetables to Irradiation below 1 K Gy

Low	Fruits	Avocado, grape, lemon, lime, olive
	Vegetable	Cucumber, green bean, pepper, summer squash, leafy vegetables, broccoli, cauliflower
Medium	Fruits	Apricot, banana, cherimoya, fig, grapefruit, loquat, litchi, orange, passion fruit, pear, pineapple, plum
High	Fruits	Apple, cherry, date, guava, muskmelon, nectarine, papaya, peach, rumbutan, raspberry, strawberry
	Vegetable	Tomato

Source: Kader 1986.

5. Chemical Control

Various chemicals are applied to fruits and vegetables in order to control postharvest diseases and pest infestations.

Methods of Application of Chemicals

a. *Dipping* : The commodity is immersed in water containing appropriate concentration of chemical which is toxic to the pathogen. However, the concentration of chemical should not be toxic to the fruit/ vegetable and should not endanger public health. For improving the efficacy of the dip treatment and better surface coating some wetting agents may also be added. The effectiveness of the fungicidal solution may also be enhanced by heating the water in which the fruit is being dipped. 500 ppm of benomyl in water at 50-55°C, for 2 to 15 min is effective for controlling anthracnose in mango without damaging the fruit.

b. *Cascade application* : Fruit / vegetables are passed below a shower of diluted chemical.

c. *Electrostatic sprays*: Applying the chemical as a spray but producing very fine particles and then charging them in an electrostatic sprayer so that they readily stick to the commodity underneath them. The fine droplets of chemical solution have same charge and thus they repel each other and are attracted towards earth during field sprays.

d. *Dusting*: Active chemical is diluted with an inert powder i.e. talc for uniform application and reduce wastage.

e. *Fumigation*: Sulphur dioxide fumes are used for controlling postharvest diseases in grapes.

f. *Chemical pads*: Paper pads impregnated chemical are used for wrapping the fruits and vegetables and control postharvest diseases.

Some Common Chemicals Recommended for Postharvest Fungal Disease Control

Chemical	Pathogens
Benomyl	*Botrytis , Colletotrichum, Penicillium, Sclerotinia*
Captan	*Botrytis , Sclerotinia*
Carbendazim	*Botrytis , Colletotrichum, Penicillium, Sclerotinia*
Mancozeb	*Pithium, Phytopthora*
Sodium carbonate	*Penicillium*
Thiram	*Botrytis, Cladosporium*
Ziram	*Alternaria*
Sorbic acid	*Alternaria, Cladosporium*
Sulphur dioxide	*Botrytis*

6. Vapour Heat Treatment (VHT)

This was developed to control infections of fruit flies in fruits. The treatment consists of stacking the fruits in boxes in a room which is heated and humidified by injection of steam. The temperature and exposure time may be adjusted depending upon the stage at which the fly is to be killed i.e. egg, larvae, pupa or adult. The most difficult stage to control by VHT is larval stage as the insect goes further into the fruit and away from the surface thus requiring high temperatures for short time. Generally the treatment of citrus, papaya, mango or pineapples may be given at 43° C in saturated air for 8 hrs followed by maintaining the temperature for further 6 hrs.

7. Degreening

Maturation of fruits is indicated by change in skin colour from green to yellow or orange. For degreening, oranges are kept at 27° C and 85-92% RH with ethylene gas @ 20-30 ppm for 24-48 hrs.

PROTECTIVE COATINGS

Waxing/ Coating : Process of applying wax on the surface of commodity by spraying, dip or immersion, brushing, fogging or foaming. Some fruits develop natural fruit wax on their surface at the time of maturity. i.e. plum, apple, citrus, grapes etc. This has its role in reducing water loss from the commodity and thus reducing shriveling and weight loss. While handling, care is taken to touch the fruits as minimum as possible to retain as much of the natural wax (also called *bloom*) on the fruit.

Advantages of Waxing

- Improve appearance of fruit
- Reduce moisture loss by 30-50% and retards wilting/ shriveling
- Heals minor injuries
- Protects fruits from minor infections
- Provides modified atmosphere and increase shelf life
- Acts a carrier for various chemicals etc

Types of waxes: Paraffin wax, carnauba wax, bee wax etc.

Examples of Some Commercial Formulations

- Tal-Prolong
- Semper Fresh for apple
- Frutox
- Waxol, Ban seal
- Nipro fruit wax for apple, and citrus
- Ban seel for banana
- Nu-coat flo for citrus
- Brilloshine L for apples, avocado, melons

Pre-Cooling

Pre-cooling in the prompt cooling of the commodity (fruits & vegetables) immediately after harvest (generally within 24 hrs of harvest), before shipment or storage which aims at removal of field heat.

Rate of cooling : It depends on

- Initial product temperature
- Rate of flow of cooling media around the commodity
- Temperature difference between produce & cooling media
- Thermal conductivity of produce

Cooling Methods

Room cooling	All fruits and vegetables (Banana, beans, cabbage, coconut, garlic, ginger, lime, lemon, onion, orange, cucumber, pineapple, potato, pumpkin, radish, sweet potato, tomato, watermelon etc.)
Forced air cooling	Fruits and fruit type vegetables (banana, Barbados cherry, berries, brussels sprouts, cucumber, eggplant, fig, ginger, grape, guava, litchi, mango, kiwifruit, mushroom, okra, oranges, papaya, passion fruit, bell pepper, pineapple, persimmon, pomegranate, quince, sapota, strawberry, tomato, tubers, cauliflower etc.)

Hydro cooling	Stem and leafy vegetables, fruits (artichoke, asparagus, beet, broccoli, brussels sprouts, cantaloupe, carrot, cauliflower, celery, Chinese cabbage, cucumber, brinjal, green onion, kiwifruit, leek, knoll-khol, orange, parsley, parsnip, peas, pomegranate, radish, spinach, rhubarb, swiss chard, summer squash, water cress etc.)
Ice cooling (Package icing)	Roots, stems, flower type vegetables (endive, broccoli, Brussels sprouts, carrot, spring onions, Chinese cabbage, leek, parsley, snow peas, spinach, swiss chard, sweet corn etc.)
Vacuum cooling	Some stem, leafy and flower type vegetables (endive, brussels sprouts, carrot, cauliflower, Chinese cabbage, celery, leek, mushrooms, lima bean, spinach, sweet corn, swiss chard etc.)
Transit icing	All fruits and vegetables

Note: Some fungicides may be mixed in water during hydro-cooling to reduce decay incidence

Weight loss during forced air cooling can be reduced by maintaining high (95%) relative humidity in the pre-cooling chamber

Washing, Cleaning and Trimming

Washing is done in plain water or chlorinated water or fungicides i.e. diphenyl amine (0.1-0.25%) and Ethoxyquin (0.2-0.5%) as postharvest dip

Sorting and Grading : This may be done manually or mechanically. The fruits after harvesting are graded on the basis of their colour, size and weight and sorted for freeness from damage/diseases. The different standard grades of some of the fruits are as follows:

a. Size Grades of Apple

Grade	Min. Dia (mm)
Super Large	85
Extra Large	80
Large	75
Medium	70
Small	65
Extra Small	60
Pitoo	55

b. Size Grades of Apricot

Grade	Min. Dia (mm)
Special	42
Grade I	36-42
Grade II	< 36

c. Size Grades for Malta Oranges

Grade	Min dia. (inches)	Min. dia. (cm)
Extra special	3.25	8.255
Special	3.00	7.620
Good	2.75	6.985
A	2.50	6.350

d. Size Grades for Nagpur Oranges

The fruit are inspected and unripe, immature, undersized, damaged or decayed fruits are discarded. For local markets, the citrus fruits are graded as per the size into small, medium and large grades. The differences between categories will depend on the type of fruit.

Grade	Size (mm)	No. of Fruits/10 Kg Packing
Small	50-55	115
	50-60	98
Medium	60-65	84
	65-70	76
Large	70-75	64
	> 75	-

e. Size Grades for Pomegranate

Grades	Fruit weight (g)	Fruit Characteristics
Super	> 750	Attractive, very large, dark red coloured, without blemishes
King	500-750	Attractive, large without blemishes
Queen	400-500	Large medium, attractive without blemishes
Prince	300-400	Medium attractive, without blemishes
12-A	250-300	Small fruits, may have 1-2 spots
12-B	< 250	Very small size

f. Size Grades for Pineapple

Fruits are graded according to size, shape, maturity, and sorted for freedom from diseases and blemishes. The cut surface is treated with a suitable fungicide to control decay.

Grade	Fruit Weight (g)
A	> 1500
B	Between 1100 - 1500
C	Between 800 - 1100
D	< 800
Baby	550 (Approx.)

g. Size Grades for Onion

Grade	Bulb dia. (cm)
Extra Large	> 6
Medium	4-6
Small	2-4

h. Size Grades for Mango

The export quality mangoes are categorized into three grades according to the fruit weight.

Grade	Fruit weight (g)
Category I	200-250
Category II	251-300
Category III	301-350

i. Size Grades for Tomato

Based on the fruit size three grades are formed.

Grade	Fruit weight (g)
Small	< 100
Medium	100-255
Large	> 255

j. Size Grades for Litchi

Grade	dia (mm)
Extra class	33
Class I	28
Class II	23

k. Size Grades for Guava

Grade	Weight (g)	Dia. (mm)
A	> 350	> 95
B	251-350	86-95
C	201-250	76-85
D	151-200	66-75
E	101-150	54-65
F	61-100	43-53

l. Size Grades for Cabbage

Grade	Wight (g)
A	200-600
B	601-1200
C	> 1200

m. Size Grades for Papaya

Grade	Wight (g)
A	200-300
B	301-400
C	401-500
D	501-600
E	601-700
F	701-800
G	801-1100
H	1101-1500
I	1501-2000
J	> 2001

Difference Between Sorting and Grading

Sorting	Grading
1. Undesirable type of fruits such as diseased, damaged, deformed, are removed	1. The fruits are categorized according to difference in their weight, size, colour, maturity etc.
2. Done to reduce spread of infection to other fruits	2. Done to fetch better price in the market.

Curing

Curing is a technique where the commodity is left in the field itself in a heap under shade for few days. It is an effective operation to reduce water loss during storage from some of the vegetables, viz., onion, garlic, sweet potato etc. grown underground.

In case of onion *curing* is a drying process intended to dry off the necks and 2-3 outer scales of the bulbs to prevent the loss of moisture and the attack by decay during storage. The outermost layer, which may be contaminated with soil, usually falls away easily on curing. The dry under-layer should have an attractive appearance. Onions are considered cured generally when they have lost 3 to 5% of their weight. Generally, they are dried in the field by stacking in a warm, covered area with good ventilation. However, in cool and moist climates, onions are cured with artificial heat blown through a duct at 30°C. Onions can also be cured by tying the tops of the bulbs in bunches and hanging them on a horizontal support of pole, wire etc. in a well-ventilated and shaded place. Curing in shade improves bulb colour.

The essential conditions during curing are:

- Heat (~ 30°C)
- Good ventilation
- Low humidity

Packaging

Packaging of fruits is primarily undertaken to assemble the produce in convenient units for storage, transportation, marketing and distribution. There are various types of packaging material used for handling fresh fruits and vegetables. The packing material and methods of some of the important fruits have been described hereunder.

a. Pomegranate

The graded fruits are wrapped in paper and packed in corrugated fibreboard (CFB) boxes containing paper shreds as cushioning material. Brown coloured 3 ply CFB boxes are used for local market while, white coloured 5 ply CFB boxes are used for distant markets.

Packing of Pomegranate

Grade (s)	Size of CFB box (cm)
Super and King	32.5 x 22.5 x 10
Queen	37.5 x 27.5 x 10
Prince and 12 A	35 x 25 x 10

b. Citrus

Fruits are packed in sacks, bags, bamboo baskets, wooden boxes or CFB boxes for sending to the markets. For export of Nagpur mandarin, usually 2 piece, telescopic, CFB boxes of three ply or five ply is used. Normally a box size of 49.5 x 29.5 x 17.5 cm having 5% ventilation with 10 kg capacity is recommended. To immobilize the fruits inside the box, three ply wax treated dividers having ventilation holes are used. In India, lime and lemons are packed in jute bags for domestic marketing.

c. Grapes

Grapes for local market are packed in ventilated CFB boxes lined with newsprint paper having 2-4 kg capacity. Fine paper shreds or fine hay is spread at the bottom and top of the box for the purpose of cushioning. Open flaps of the box are fixed by an adhesive tape. Table grapes for exports are packed in 5 ply corrugated boxes of size 500 x 300 mm and 5 Kg capacity. The graded bunches (1-2) weighing between 350-650 g are packed in small and thin polythene pouches are then packed in 5 Kg, 5 ply CFB boxes with bubble sheet for cushioning.

d. Litchi

Litchi should be packed as quickly as possible after harvesting as its quality deteriorates markedly, if the fruits are exposed to Sun even for a few hours. For domestic markets litchi is usually packed in clusters, in small bamboo baskets or wooden crates lined with litchi leaves or other soft packing material as paper shavings, wood-wool, etc. Box should have few holes for ventilation and rope handles on either side for lifting the box.

e. Mango

Larger sized mango fruits take 2-4 days more time in ripening than smaller ones. Therefore, proper grading of fruits is necessary for uniform ripening. For exports, the pedicel of the fruits is cut approximately at 1 cm from the fruit with the help of sharp scissors and fruits are kept up side down in special knitted pallets, for two hours so that the latex flows out from the fruit completely. Avoid dropping latex on the fruit.

Generally mango is harvested at an early stage and the fruits do not ripe uniformly without any ripening aid. Fruits are ripened uniformly by dipping in 750 ppm etherel (1.8 ml/litre) in hot water at 52±2°C for 5 minutes within 4-8 days under ambient conditions. Mature fruits require lower doses of etherel for uniform ripening.

Wooden boxes or CFB boxes of 5 kg and 10 kg capacity are commonly used for packaging and transportation of mango fruits. Paper scraps, newspapers, etc., are used as cushioning material for the packaging of fruits. Wrapping of fruits individually (Unipack) with newspaper or tissue paper and packing in honeycomb nets helps in optimum ripening with reduced bruising and spoilage.

f. Pineapple

For local markets, fruits are packed in bamboo baskets (20-25 kg) lined with paddy-straw. The first layer of fruits is arranged to stand on their stumps and second layer on the crowns of the first layer fruits. For distant markets, fruit are wrapped individually with paddy straw and then packed. For export vertically or horizontally arranged pineapples are packed into fibreboard or wood containers. The interspace between the fruits is filled with straw.

g. Apple

Traditionally apple, after packing the fruits individually in paper, were packed in wooden boxes with newspaper lining. But now the fruits are packed in specially designed trays according to the size grade of fruits and these trays are then placed in telescopic CFB cartons of 20-23 Kg capacity and tied with a strong polystrap. The height of the telescopic CFB carton can be increased or decreased according to the size of fruits and number of layers placed in one box. An inverted tray is placed over the top layer of fruits before closing and sealing the box.

h. Banana

For packing and transportation the bunches are padded with banana leaves. A fungicidal solution/paste is applied to cut ends to prevent stem end rot. When the paste dries, bunches are stacked vertically or horizontally in trucks or railway wagons with wilted or dry banana leaves. For exports, banana hands and clusters of the bananas are packed in corrugated boxes with perforated polyethylene liners. Curved side of the hands is kept facing upwards making sure that the crown of the upper hands do not damage the lower fruits.

i. Onion and Garlic

Onions and garlic are packed in 40 kg open mesh jute (hessian) bags (having 200-300 g weight) for transportion in domestic markets. For exports, graded onions are packed in 5-25 kg capacity open mesh jute bags or 14-15 kg capacity wooden baskets. Nylon net bags, having good durability and giving attractive appearance to the produce are also being used for reducing loss due to good ventilation.

j. Tomato

Earlier tomato fruits were packed in bamboo baskets or wooden baskets lined with news papers for local markets. But, now 20-22 Kg capacity plastic crates are commonly used. These plastic crates besides having good ventilation, rigidity and their water resistant quality can also be stacked conveniently one over the other. For exports, tomato are packed in cardboard telescopic boxes (<15 Kg capacity).

k. Stone Fruits (peach, plum, apricot, green almonds)

Stone fruits are packed in small packs of 4-6 Kg capacity in CFB or cardboard boxes having proper vertiliation for domestic markets due to their highly perishable nature.

l. Capsicum

Capsicum is packed in plastic crates for domestic markets. Packing in gunny bags results in huge bruising damage and poor stacking strength thus causing more losses.

m. Cabbage, Cauliflower, Pea

These vegetables are generally packed directly in 25-30 kg capacity open mesh jute bags for transportation and domestic marketing. In cabbage and cauliflower, the outer leaves are retained and not removed for cushioning.

Points to Remember

- √ Harvesting is detaching a commodity from the point of its origin.
- √ Goal of harvesting is to gather the commodity at optimum stage of maturity depending upon the purpose with minimum of damage and loss, in as minimum time as possible and at the lowest possible cost.
- √ Mechanical harvesting is very useful for rapid harvesting of a particular crop at low cost.
- √ Maturity is a stage of fruit growth preceding ripening or stage at which commodity has attained sufficient physiological development such that after harvesting at that stage and adequate postharvest handling (including controlled ripening if required), its quality would be above the minimum acceptable to the ultimate consumer.
- √ Physiological maturity is the stage at which a plant or a plant part continues ontogeny even if detached.
- √ Horticultural maturity is the stage at which a plant or a plant part possesses all the pre-requisites for utilization by the ultimate consumer for a particular purpose.
- √ Maturity indices are the factors on the basis of which it is decided whether the horticultural commodity is ready for harvesting or not. A single maturity index is not considered to be reliable.
- √ Freezing injury occurs when a commodity is held below its freezing temperature
- √ Chilling injury occurs in subtropical or tropical fruit crops when a they are held above their freezing point but below their optimum storage temperature (5-15° C).
- √ Heat injury occurs when the commodity is exposed to direct sunlight or excessively high temperature.
- √ Bitter pit and blossom end rot occur due to the deficiency of calcium.
- √ Waxing/ Coating is the process of applying wax on the surface of commodity by spraying, dip or immersion, brushing, fogging

or foaming to predominantly to prevent weight loss and improve appearance.

- √ Curing is the technique where the commodity is left in the field itself in a heap under shade for few days. It is an effective operation to reduce water loss during storage from hardy vegetables viz., onion, garlic, sweet potato etc.
- √ Packaging of fruits is primarily undertaken to assemble the produce in convenient units for storage, transporation, marketing and distribution.

4

Storage of Fruits and Vegetables

Storage : It can be defined as keeping the commodity fresh or processed in safe condition with minimum of deteriorative changes for later use.

Reasons of Deterioration During Storage

1. Physical

a. Damage caused by insects, rodents, animals etc.
b. Weight loss and shriveling
c. Bruising during handling

2. Physiological

a. Physiological disorders, i.e. freezing injury, chilling injury etc.
b. Fibre development (beans)

c. Greening (potatoes), sweetness loss (pea)
d. Sprouting (root, shoot growth)
e. Seed germination

3. Chemical

a. Loss of carbohydrates and vitamins
b. Development of off flavours
c. Changes in surface colour (green vegetables)

4. Microbial

a. Postharvest diseases caused by pathogens, rotting, fermentation etc.

Goals of Storage

- Slow down biological activity i.e. respiration rate, ethylene evolution etc.
- Reduce product drying and moisture loss (shriveling etc)
- Reduce growth of microorganisms and reduce pathogenic damage (rotting, diseases etc.)
- Avoid physiological disorders (freezing injury, chilling injury, heat injury, storage disorders etc.)
- Reduce physical damage (due to rodents, animals etc.)

Storage Considerations

1. *Temperature*: Van't Hoff Quotient (Q_{10}) - The rate of deteriorative reactions doubles for each 10° C rise in temp. Therefore, the storage temperature should be optimum. Higher temperature cause increase in respiration and lower temperature may cause chilling or freezing injury.
2. *Relative Humidity* : Relative humidity should be at the optimum level. Lower humidity levels may cause shrinkage and weight loss. Generally for most of the fruits and vegetables relative humidity levels of 85-95% are

optimum except some commodities i.e. onion, garlic, potato etc. where humidity levels of 65-75% are required. Higher humidity in such commodities may cause rotting.

3. *Atmospheric Composition*: Composition of gases affects the storage life of any commodity in a particular set of conditions of temperature and humidity. More the availability of oxygen higher shall be the respiration rates (upto a specific level). Absence of oxygen may initiate anaerobic respiration, leading to the formation of alcohol. Presence of ethylene in the storage atmosphere also favours respiration, although the pattern of respiration in the presence of same levels of ethylene may differ in climacteric and non-climacteric fruits.

General Recommended Storage Conditions for Vegetables

Fruit/vegetable	Temperature (°C)	Relative humidity (%)	Approx. storage life
Asparagus	0	95	3 - 4 weeks
Beans green	4 - 7.2	90 - 95	1 - 2 weeks
Bitter gourd	0.6 - 1.7	85 - 90	1 month
Brinjal	8 - 12	90 - 95	1 - 3 weeks
Broccoli	0	95	1.5 - 2 weeks
Cabbage	0 - 1.7	92 - 98	3 - 6 weeks
Carrot	0	95	20 - 24 weeks
Cauliflower	0 - 1.7	85-98	3 - 7 weeks
Chilly	7	90-95	2-3 weeks
Colocasia	11.1 - 12.8	85-90	21 weeks
Cucumber	10 - 13	92-95	1.5 - 2 weeks
Garlic	0	65 - 70	6 - 9 months
Ginger	7.2 - 13	65 -75	4 - 6 months
Lettuce leaves	0	95-98	1 - 3 weeks
Mushroom	0	95	3 - 4 days
Muskmelon	1.7 - 7.2	85 -90	1.5 - 4.5 weeks
Okra	8 - 9	90	2 weeks

Onion	0	70 - 75	5 -6 months
Pea	0	88 - 98	1-3 weeks
Potato	3 - 5	85	5 – 10 months
Pumpkin	1.7 – 11.6	70 - 75	6 – 9 months
Radish	0	88 – 92	3 – 5 weeks
Spinach	0	95 -100	10 -14 days
Tomato	7.2 - 10	85 - 90	1 - 4.5 weeks
Turnip	0	90 - 95	8 – 16 weeks
Watermelon	7.2 – 15.6	80 -90	2 weeks

General Recommended Storage Conditions for Fruits

Apple	0 – 4	80 – 90	4 – 8 months
Apricot	0 – 2	80 – 85	1 – 2 weeks
Banana	12 – 13	80 – 90	1 – 2 weeks
Cherry	0 – 2	85-90	2 weeks
Dates	7 - 8	85- 90	2 weeks
Fig	0 - 2	85 - 90	4 weeks
Grape	-1.1 - 2	80-95	3-8 weeks
Guava	7.2 - 10	85- 90	2 – 3 weeks
Jackfruit	11 - 13	85 - 90	1.5 months
Kiwifruit	-0.5 - 0	90 - 95	3 – 5 months
Lemon	8 - 10	85 -90	1 - 6 months
Lime	8-10	85 -90	3 - 6 weeks
Litchi	0 – 2.1	85-95	3-10 weeks
Mandarin	5 - 8	85 - 90	10 -14 weeks
Mango	8 - 12.8	85 -90	2 – 7 weeks
Papaya	7.2 – 10	80-90	1-2 weeks
Passion fruit	7 - 8	80 - 85	4 – 5 weeks
Pear	0 – 1	85 -90	3 – 6 months
Peach	0 – 3	85 - 90	2 – 4 weeks
Persimmon	0 - 2	85 - 90	7 weeks
Pineapple	8 - 13	85 - 90	3 – 6 weeks
Plum	0 – 2	85 - 90	4 – 8 weeks
Pomegranate	0 – 2	80 - 90	2 – 6 weeks
Sapota (Chiku)	3 – 4	85 - 90	6 – 8
Strawberry	0 – 2	80 -95	5-7 days

Cool Chain Concept : Handling of the produce at optimum (low) temperature and relative humidity, conditions immediately after harvest till commodity reaches the consumer, aimed to reduce losses, extend storage life and retain fresh like quality. Cool chain is also sometimes called cold chain.

General Steps for Cold/cool Chain of Horticultural Commodities

1. Harvesting during cooler part of day for reducing field heat
2. Pre-cooling
3. Cold storage at field conditions
4. Refrigerated transport from field to destination market / airport
5. Palletization in cold room for air transport
6. Cold storage at destination / marketing centres
7. Cold storage by the consumer before consumption

Storage Systems

1. In situ

The commodity is left in the field itself (*in* situ) for some time even after maturity and harvesting is delayed till its requirement in the market. eg. commodities grown underneath the soil.

Disadvantages

- Field remains occupied
- Chances of damage due to insect-pests and diseases.
- Development of undesirable fibre and starch in some commodities.

2. Pits and Trenches

Pits or trenches (generally 1-1.5 m deep) are dug at a raised, shady and cooler place in the fields. These are then lined with straw and commodity is put for storage. On the top also a layer of straw is placed to cover the commodity completely followed by covering with wooden planks or soil. A hole should be kept for ventilation and to avoid rotting. Suitable for storing ginger.

Disadvantages

- Can not maintain high humidity so not suitable for fruits leafy vegetables demanding high humidity
- The stored commodity cannot be examined frequently for any incidence of rotting etc.

3. Clamps

A layer of straw (15-20 cm thick) is laid on the surface of soil at a raised place free from water logging. Potatoes are placed in a conical heap. The top of the heap is also covered with straw (15-20 cm thick) and finally covered with soil. This storage is used for potato, cassava etc.

Disadvantages

- Desiccation due to low relative humidity.
- Large heaps may result in more incidence of rotting

4. Cellars

Storage of commodities in underground or partially underground rooms over shelves for maintaining proper ventilation. These rooms are almost insulated to outside temperatures and are cooled by opening the doors during night time.

Disadvantages

- Desiccation due to low relative humidity.

5. Evaporative Coolers

a. Zero Energy Cool Chamber (ZECC)

ZECC is a low cost storage structure suitable for short duration storage of fruits and vegetables. There is no need of any power source i.e. electricity, diesel, petrol etc. for cooling, thus, the name zero energy cool chamber.

Principle : ZECC works on the principle that "evaporation causes cooling". When water is exposed to atmosphere, it evaporates and the energy (heat) in the surroundings is used up in converting water from liquid state to vapours which results in cooling of the surroundings. The vapours are then taken away by air (wind).

The rate of evaporation therefore would depend on

i. *Temperature of atmosphere*: higher the temperature more would be the evaporation and more the cooling
ii. *Wind velocity* : Higher the wind velocity better the evaporation
iii. *Relative humidity* : The relative humidity in present context would be the moisture carrying capacity of air i.e. saturation of air with vapours. Lesser the relative humidity higher the rate of evaporation. When air is saturated with moisture, it has no more capacity to carry the vapours. Thus, even if the temperatures are high during rainy season, the drying of clothes or any material is difficult.

Design

a. Floor space- brick layer
b. Outer dimensions – 170-180 cm (L) x 115-125 (B) cm x 68 -72 (H) cm. Length may be increased for higher storage capacity but breadth and height should not be increased as this may hinder working in the chamber and reduce cooling capacity. More height would make placing and removing the crates of fruits and vegetables difficult and more breadth would reduce the cooling rate of crates placed in the centre.
c. Distance between two brick walls – 7.0 - 8.0 cm
d. Cavity – filled with river sand
e. Cover – made of gunny cloth in bamboo frame
f. Location – Shady place with lots of aeration and preferably should be closer to water source

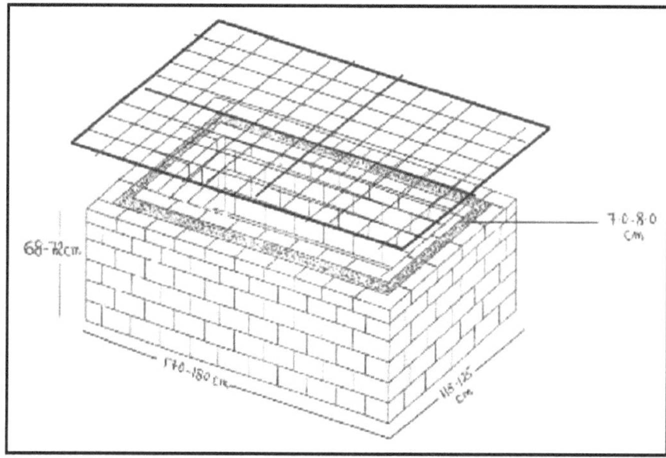

Line Diagram of Zero Energy Cool Chamber

Operation: Whole structure is made wet by sprinkling water. Fruits and vegetables are stored in plastic crates. Max. / Min. temperature thermometers and dry and wet bulb thermometers should be placed.

Storage life - Causes a temperature drop of 10-18°C and storage life increases by 2-3 times.

View of ZECC Under Operation

Suitable for - almost all fruits and vegetables including apple, mango, plum. apricot, pear, guava, citrus, and vegetables like capsicum, tomato, cucumber, bottle guard, leafy vegetables, spinach, green chillies, bitter gourd, peas etc.

Not suitable for - onion, garlic, ginger, potato etc. These crops require comparatively lesser relative humidity (65-80 %) for their storage, and in ZECC relative humidity of 90-95 % is maintained. This may lead to incidence of rotting in those commodities which require low RH for their storage.

Precautions

1. While making the chamber do not plaster it with cement for better evaporation and cooling.
2. Never construct the chamber with stones. Use bricks as they have better water absorption capacity and thus better evaporation and cooling.
3. Water the chamber regularly at least 2-3 times a day
4. Always store the fruits or vegetables in plastic crates / trays only
5. Do not let the water fall directly over the fruits or vegetables
6. Take due care of the commodities stored otherwise they may be spoilt by rodents

Limitations: This chamber does not work properly under two conditions

1. If it is not properly watered because it is entirely based on the principle that evaporation causes cooling.
2. During the rainy season i.e. July-August, when the rate of evaporation slows down very much and the ultimate principle of its functioning is defeated.

b. Walk- in Evaporative Cool Chamber

This is an enlarged version of ZECC but it requires electricity to run exhaust fans for better air circulation and cooling. This chamber is about 2 m in height and a person can straight way walk into the chamber and work. The fruits and vegetables are stored in plastic crates which can easily be stacked one over the other. Water continuously trickles in cooling pads provided on the rear side of the chamber. Insulation is made by making the walls double as in case of ZECC.

6. Refrigerated Storage

Refrigeration is the process of removing heat from an enclosed space, or from a substance or commodity. The primary purpose of refrigeration is lowering the temperature of the enclosed space or substance or commodity and then maintaining that lower temperature.

Cooling refers to any natural or artificial process by which heat is dissipated.

Cryogenics : Process of artificially producing extremely cold temperatures by using cryogenic refrigerants.

Cold is absence of heat. In order to decrease temperature, one must "remove heat", rather than "adding cold".

Refrigeration Cycle

The basis of mechanical refrigeration is the fact that at different pressures the condensing temperatures of vapours are different. As the pressure increases condensing temperatures also increase and *vice versa*.

This process of evaporation of refrigerent at a low pressure and low temperature, followed by compression and condensation at high (around atmospheric) temperature and high pressure, is the *refrigeration cycle*. The high-pressure liquid then passes through a nozzle from the condenser to the evaporator at low pressure, and the cycle continues.

In *evaporator* the pressure is low enough so that evaporation of the refrigerant liquid to a gas occurs at low temperature (depending on the cooling requirements of product). This low temperature might be, say -18°C and the corresponding low pressures for ammonia would be 229 kPa absolute and for tetrafluoroethane (also known as refrigerant 134a) 144 kPa. Evaporation then occurs at such low temperature and this extracts the latent heat of vaporization for the refrigerant from the surroundings of the evaporator. The low pressure necessary for the evaporation at the required temperature is maintained by the suction of the compressor.

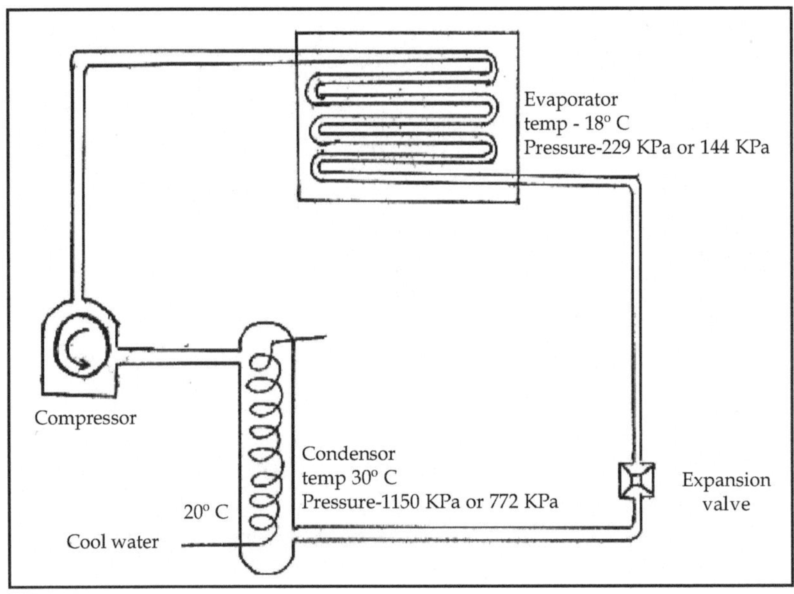

Fig. Refrigeration Circuit

The rest of the process cycle is meant merely to return the refrigerant to the evaporator and to continue the cycle. First, the vapours are sucked into a compressor (which is just like a gas pump) which increases its pressure to exhaust and release it at the higher pressure to the condensers. The higher pressure might be, say 1150 kPa (pressures are absolute pressures) for ammonia, or 772 kPa for refrigerant 134a, and it is determined by the temperature at which cooling water or air is available to cool the condensers.

The refrigerant must be condensed, giving up its latent heat of vaporization to some cooling medium to complete the cycle. This is carried out in a condenser, which is a heat exchanger cooled generally by water or air. The condensing temperature, corresponding to the above high pressures, is about 30°C for which cooling water at about 20° C, could be used. By adjusting the high and low pressures, the condensing and evaporating temperatures can be selected as required.

Refrigeration Load

The most common unit used to quantify refrigeration load is the refrigeration ton and refrigeration tonne.

One *ton of refrigeration* is defined as the energy removed from one ton (2,000 lbs) of water so that it freezes within 24 hours at 0° C (32 °F). It is equivalent to 2,88,000 Btu/day or 12,000 Btu/h.

Latent heat of ice (i.e., heat of fusion)
= 333.55 kJ/kg ≃ 144 Btu/lb

One short ton = 2000 lb

Heat removed = (2000) x (144)/24 hr =2,88,000 Btu/24 hr
= 12,000 Btu/hr = 200 Btu/min

1 ton refrigeration = 200 Btu/min = 3.517 kJ/s = 3.517 KW

One *tonne of refrigeration* is defined as the energy removed from one metric tonne (1000 Kg) of water so that it freezes within 24 hours at 0° C (32 °F).

1 tonne of refrigeration = 13,898 kJ/h = 3.861 kW

Therefore, *1 tonne of refrigeration is about 10% larger than 1 ton of refrigeration.*

Most residential air conditioning units range in capacity from about 1 to 5 tons of refrigeration

Determining the refrigeration load requires quantification of field heat and heat of respiration of the produce, conductive heat gain, convective heat gain, equipment load, and service and defrost factors of the facility.

Field Heat : Amount of cooling necessary to reduce the produce from harvest temperature (more specifically entry temperature into cold storage) to the safe storage temperature within a specified time period. The hotter the produce at the time of entry into cold storage, the more energy would be needed to reduce the temperature to safe limits and the more time required to operate the equipment.

Heat of Respiration : Energy released by the produce as it respires during storage. The warmer the produce, the more heat of respiration generated.

Conductive Heat Gain : Heat gained by conduction through the building floor, walls, roof, ceiling etc. It is directly related to the quality of insulation installed in the cold storage. The better the quality of insulation the lesser the conductive heat gain.

Convective Heat Gain : Heat that enters the cold storage during the mixing of outside air with the cool inside environment during opening of doors etc. This load is directly related to the number of doors and their time of opening. The more the number of doors in the cold storage, the more is the possibility of mixing air currents. The amount of heat gained increases with the duration of time and periods for which the doors remain open.

Equipment Load : It is the heat generated by equipments operating in the room i.e fans and lights. Do not put excess number of lights inside the cold storage. Lighting should be, just enough to identify produce and labels clearly, and to allow safe working.

Service and Defrost Factors : The service factor are brief periods of unusually hot weather, loading rates that temporarily exceed those anticipated, or other unexpected conditions. The defrost factor is the time lost during coil defrost.

Exercise : Calculate the Refrigeration Requirements During Cold Storage of Apple at 1°C

Suppose that the apple storage is having outer dimensions of 15 x 15 x 5 m^3. The fruits are harvested at 32° C and 55% RH with the fruit internal temperature being 23° C at harvest. Fruits are stored in bins of capacity 500 Kg/ bin and the bin weight being 60 kg/bin. During day 1 the temperatures are brought from 23°C to 6°C and then to 1° C during the day 2. Air changes due to door opening during cooling is 6 times and specific heat of fruit and bins is 0.86 and 0.50 respectively. Heat load in lowering air at 32°C and 55% RH to 1°C is 74.5 kJ/m^3 and miscellaneous loads include lights (2,500 W/h), fans (3 hp), two electric forklift trucks (37,000 kJ each for 8 h) and two men working for 8h (1,000 Kj/h each). Calculate the refrigeration requirements in KJ/24 h.

Storage dimensions outer (l x b x h) m^3	15 x 15 x 5 m
Outside surface area, including floor	750 m^2
Storage dimensions inner (l x b x h) m^3	14.7 x 14.7 x 4.7 m
Volume (inner)	1016 m^3
Insulation	7.5 cm of polyurethane Conductivity value (k) = 1.3 kj/ m^2/cm thickness /°C Coefficient of transmission (U) =1.1 kJ/h/m^2/°c
Ambient conditions at harvest	32° C and 55% relative humidity (RH)

contd...

contd...

Fruit temperature	At harvest 23^0C In storage 1^0C
Storage capacity	600 bins @ 500 kg fruits/bin Total = 3,00,000 kg of fruit
Bin weight	60 kg Total weight if bins 36,000 kg
Loading rate	200 bins i.e. 1,00,000 kg fruit/day
Days reqd. to fill	3 days
Cooling rate	I day 23^0C to 6^0C II day 6^0C to 1^0C
Air changes due to door opening during cooling	6 per day
Air changes due to door opening during storage	1.8 per day
Specific heat (sp.ht.)	For fruits say 0.86 For bins say 0.50
Heat load in lowering air at 32^0C and 55% RH to 1^0C	74.5 kJ/m^3
Miscellaneous heat loads	Lights, 2,500 W/h Fans, 3 hp Two electric forklift trucks, 37,000 kJ each for 8 h. Two men working for 8h, 1,000 Kj/h each

Specific Heat = 0.008 x % water in food + 0.20

Specific heat is defined as the ratio of heat required to raise the temperature of a given weight of material i.e. fruits / vegetables etc. to that required to cause an equivalent rise in the same weight of pure water.

1. Load during cooling and filling storage	**KJ/24 h**
(Temperature difference (TD) =32° C to 1°C =31°C	
(assume 31°C TD on all the surfaces)	
Building transmission load:	6,13,800
Area ($750m^2$) X U (1.1Kj) X TD(31°C) X HOURS(24)=	
Air change load from door openings:	4,54,152
Volume (1016 m^3) X heat load (74.5 kJ) X air changes (6) =	
Product load:	
Product cooling (field heat, or sensible heat, removal) –	
First day	
Fruit wt. (1,00,000kg) X sp.ht. (0.86) X TD(17° C to 6° C) X kJ factor (4.186) =	61,19,932
Bin wt. (12,000 kg) X sp.ht. (0.5) X TD(17°Cto 6°C) X kJ factor (4.186) =	4,26,972
Second day	
Fruit wt. (1,00,000kg) X sp.ht. (0.86) X TD(5°C to 1°C) X kJ factor (4.186) =	17,99,980
Bin wt.(12,000kg) X sp.ht. (0.5) X TD(5°C to 1°C) X kJ factor (4.186) =	1,25,580
Heat of respiration during cooling (vital heat)	-
First day	
(Av. Temp., 13°C: resp. rate, 12,200 Kj /t/24h.)	
tonnes of fruit (100) x rate (12,200) =	12,20,000
Second day	
(Av. Temp., 1.7°C: resp. rate, 1,741 Kj /t/24h.)	
tonne of fruit (100) x rate (1,741) =	1,74,100
Maximum accumulated in storage before cooling completed (total fruit wt. (300,000 kg)- 2 days loading wt. (200,000 kg) = 100,000 kg, or 100 t.:	
Resp. rate at 1°C, 812 kJ/t/24h.)	
Tonne of fruit (100) x rate (812) =	81,200
Miscellaneous heat loads:	
Lights –watts (2500) Kj per watt (3.6) X hours (8) =	72,000
Fans – hp(3) x Kj per hp (3,112) x hours (24) =	2,24,064
Forklifts -2 x 37,000 kJ per forklift for (8h) =	74,000
Labor – men (2) x kJ per hour (1,000) x hours (8) =	16,000
Total heat load during cooling :	
1. Building transmission	6,13,800
2. Air change	4,54,152
3. Product cooling	84,72,464
(iv) Production respiration	14,75,300
(v) Miscellaneous	3,86,064
subtotal	1,14,01,780
Add 10% for safety	11,40,178
Total required refrigeration	**1,25,41,958**

7. Controlled Atmosphere (CA) Storage

Scientific basis: Air contains 20.9 % O_2, 78.1 % N_2, and 0.03% CO_2 alongwith other gases in traces. For normal respiration of fruits oxygen is absorbed from the surrounding atmosphere and carbon dioxide is released. If the concentration of oxygen in the storage atmosphere is reduced then, it would lead to reduction in the respiration rate and thus longer storage life of fruits and vegetables. Therefore, *in CA storage we reduce the concentration of O_2 and increase the concentration of CO_2 in storage atmosphere to fixed levels to slow down respiration rate and ripening.*

- CA storage room is insulated properly and is gas tight
- Oxygen level is held at 0-10% and CO_2 at 0-20% (or depending upon the fruit as well as variety)
- Temperatures 0-15 °C

How Oxygen Concentration is Reduced?

Oxygen concentration is reduced by addition of nitrogen gas through cylinders into the storage atmosphere. *Atmosphere generators* can also be used to remove oxygen by burning it from the air with propane or natural gas. CO_2 is also added into the storage chamber through gas cylinders.

How CO_2 Concentration is Controlled?

a. CO_2 is flushed into the storage chamber through cylinders for a particular level.

b. CO_2 scrubbers are used for removal of excess of CO_2 from the storage atmosphere. Examples of CO_2 scrubbers are hydrated lime, aluminium calcium silicate, activated carbon etc.

Recommended CA Conditions for Fruits and Vegetables

Fruit/ vegetable	Temperature (°C)	% O_2	% CO_2
Apple, apricot, peach, plum, prunes, kiwifruit	0-5	1 - 3	0 - 5
Strawberry	0-5	10	15-20
Persimon	0-5	3-5	5-8
Nuts and dried fruits	0-25	0-1	0-100
Lemon, mango, papaya, pineapple	10-15	5	5-10 (0-5 for lemon)
Banana	12-15	2-5	2-5
Avocado	5-13	2-5	3-10
Orange	5-10	10	5
Olive	8-12	2-5	5-10
Cucumber, okra, bell pepper, tomato	8-12	3-5	0
Green onion, sweet corn	0-5	1-4	10-20
Broccoli, cabbage, cauliflower, leek, lettuce, mushroom, spinach	0-5	1-5 (or even air)	2-20

Advantages of CA Storage

a. Best control over the atmospheric conditions therefore best storage results obtained (in most cases)

b. Insects are also controlled

c. Reduced sensitivity of commodity to ethylene

Limitations of CA Storage

a. Very costly technology and is beyond reach of common farmer in India

b. Problems may arise during non-availability of gases

c. Off flavours may develop at very low O_2 concentration.

d. Irregular ripening (banana, pear, tomato etc.) and physiological disorders (black heart in potato, brown stain in lettuce) have also been reported in some cases.

8. Modified Atmosphere (MA) Storage

In MA storage also the concentration of oxygen is reduced and concentration of CO_2 is increased like in case of CA storage, but the degree of control over the concentration of gases in surrounding atmosphere of the commodity is lower in MA than CA.

In MA, normally the commodity is packed in different types of polymeric films and containers and regulation in gas concentration or the generation of MA is achieved by following methods-

1. *Commodity generated MA in sealed packages* : This is also called as *Passive MA*. First of all the commodity is packed in sealed packages or films. As a result of continuous respiration of commodity inside the package, and non-permeability or selective permeability of package to oxygen, the concentration of O_2 is reduced and concentration of CO_2 is increased automatically. Whatever, oxygen is available inside the package, would be used up in respiration and CO_2 would be released, thus generating MA.

2. *Active MA*: For generation of *active MA*, the commodity is packed and sealed and gas is pulled out slightly creating vacuum. The package atmosphere is replaced with the desired mixture of gases. This ensures faster creation of the desired atmosphere inside the package. Ethylene absorbers and CO_2 scrubbers can be kept inside the package in order to further avoid buildingup of these gases to injurious levels.

3. *Oxygen absorbers* : This may be kept in the package for removal of oxygen eg. Ferrous Oxide.

4. *CO_2 absorbers* : These may be kept for removal of excess CO_2 from the storage atmosphere eg. hydrated lime, activated charcoal, MgO
5. *Ethylene absorbers* : These may be kept for removal of excess C_2H_2 from the storage atmosphere eg $KMnO_4$ absorbed on celite, vermiculite, silica gel or alumina pallets, Ethysorb (commercial preparation).

Difference Between CA and MA Storage

CA storage	MA storage
1. High degree of control over the gas concentration	1. Lower degree of control over the gas concentration
2. Longer storage life	2. Shorter storage life
3. More expensive technology	3. Less expensive technology
4. Atmosphere is modified by adding gases through cylinders	4. MA is created either actively (addition or removal of gases) or passively (commodity generated)
5. Temperature is also maintained at a specific level during storage	5. Temperature may not be maintained.

9. Hypobaric Storage/Low Pressure Storage

In this technology fruits are stored at

Low pressure	0.2 - 0.5 atmosphere and
Temperature	59-75° F or 15-23.9°C

The commodity is stored in an air tight chamber and pressures are reduced by sucking air and creating vacuum. Humidified air is passed to avoid drying due to vacuum.

Mechanism

1. Reduced O_2 supply slows down respiration. When pressure reduces from 1 atm to 0.1 atm the effective O_2 concentration reduces from 21 to 2.1 %

2. Released ethylene is removed out of storage.
3. Volatiles are also reduced / removed - CO_2, acetaldehyde, acetic acid, esters *etc.*

Points to Remember

√ Storage can be defined as keeping the commodity fresh or processed in safe condition with minimum of deteriorative changes for later use.

√ There are 4 main reasons of deterioration during storage i.e. physical, physiological, chemical and microbial

√ To slow down biological activity, reduce drying, reduce microbial growth, avoid physiological disorders and to redue physical damage are important goals of storage

√ Temperatrue, RH and atmospheric composition are important storage considerations

√ For most of the fruits and vegetables relative humidity levels of 85-95% are optimum except some commodities i.e. onion, garlic, potato etc. where humidity levels of 65-75% are required.

√ According to Van't Hoff Quotient (Q_{10}) the rate of deteriorative reactions doubles for each 10° C rise in temp.

√ Cool Chain Concept is about handling the produce at optimum (low) temp., RH conditions immediately after harvest till commodity reaches the consumer.

√ Pits and cellars are underground storage structures while, clamps are above ground structures

√ ZECC is based on the principle that evaporation causes cooling

√ ZECC is a low cost storage structure suitable for short duration storage of fruits and vegetables. There is no need of any power source i.e. electricity, diesel, petrol etc. for cooling, thus, the name zero energy cool chamber.

√ The rate of evaporation in ZECC would depend on temperature, wind velocity and RH. Causes a temperature drop of 10-18°C and storage life increases by 2-3 times.

- √ ZECC should be constructed at a shady place with lots of aeration and preferably should be closer to water source
- √ ZECC is suitable for apple, mango, plum. apricot, pear, guava, citrus, and vegetables like capsicum, tomato, cucumber, bottle guard, leafy vegetables, spinach, green chillies, bitter gourd, peas etc. and not suitable for onion, garlic, ginger, potato etc.
- √ ZECC fails to work efficiently during the rainy season i.e. July-August, when the rate of evaporation slows down very much and the ultimate principle of its functioning is defeated.
- √ *Refrigeration* is the process of removing heat from an enclosed space, or from a substance. The primary purpose of refrigeration is lowering the temperature of the enclosed space or substance and then maintaining that lower temperature
- √ *Cooling* refers to any natural or artificial process by which heat is dissipated.
- √ *Cryogenics* : Process of artificially producing extremely cold temperatures.
- √ *Cold* is absence of heat. In order to decrease temperature, one must "remove heat", rather than "adding cold".
- √ The process of evaporation at a low pressure and corresponding low temperature, followed by compression and condensation at around atmospheric temperature and corresponding high pressure, is the *refrigeration cycle.*
- √ One *ton of refrigeration* is defined as the energy removed from one ton (2,000 lbs) of water so that it freezes within 24 hours at 0° C (32 °F). It is equivalent to 2,88,000 Btu / day or 12,000 Btu/h.
- √ 1 ton refrigeration = 200 Btu/min = 3.517 kJ/s = 3.517 KW
- √ One *tonne of refrigeration* is defined as the energy removed from one metric tonne (1000 Kg) of water so that it freezes within 24 hours at 0° C (32 °F).
- √ 1 tonne of refrigeration = 13,898 kJ/h = 3.861 kW
- √ 1 tonne of refrigeration is about 10% larger than 1 ton of refrigeration.
- √ Field Heat: Amount of cooling necessary to reduce the produce from harvest temperature (entry temperature) to the safe storage

temperature within a specified time period. The hotter the produce at the time of entry into cold storage, the more energy would be needed to reduce the temperature to safe limits and the more time required to operate the equipment.

- √ *Heat of Respiration*: Energy released by the produce as it respires during storage. The warmer the produce, the more heat of respiration generated.
- √ In CA storage we reduce the concentration of O_2 and increase the concentration of CO_2 in storage atmosphere to fixed levels to slow down respiration rate and ripening. Oxygen level is held at 0-10% and CO_2 at 0-20% (or depending upon the fruit as well as variety). Temperatures 0-15 °C.
- √ Examples of CO_2 scrubbers are hydrated lime, aluminium calcium silicate, activated carbon etc.
- √ In MA storage also the concentration of oxygen is reduced and concentration of CO_2 is increased like in case of CA storage, but the degree of control over the concentration of gases in surrounding atmosphere of the commodity is lower in MA than CA.
- √ Commodity generated MA in sealed packages is also called as *Passive MA*.
- √ For generation of active MA, the commodity is packed and sealed and gas is pulled out slightly creating vacuum. The package atmosphere is replaced with the desired mixture of gases.
- √ Oxygen absorbers may be kept in the package for removal of oxygen eg. Ferrous Oxide.
- √ CO_2 absorbers may be kept for removal of excess CO_2 from the storage atmosphere eg. hydrated lime, activated charcoal, MgO
- √ Ethylene absorbers may be kept for removal of excess C_2H_2 from the storage atmosphere eg $KMnO_4$ absorbed on celite, vermiculite, silica gel or alumina pallets, Ethysorb (commercial preparation).
- √ In hypobaric storage fruits are stored at low pressure (0.2 - 0.5 atm) & low temperature (59-75° F or 15-23.9°C).
- √ Hypobaric storage is also called as low pressure storage.

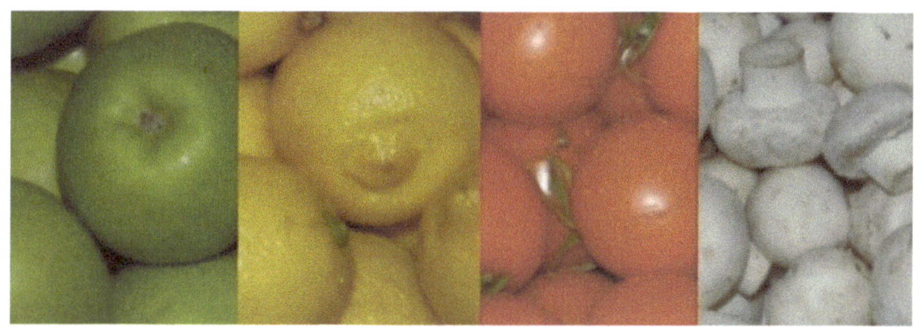

PART II
Management and Processing

5

Packaging of Fresh and Processed Food Products

Packaging : Packaging of fruits, vegetables, flowers or any other commodity is chiefly done to assemble the produce in convenient units of different sizes, shapes & types for storage, transportation, marketing and distribution. The size of units may vary depending on the requirement e.g for retail marketing smaller units are preferred and for bulk transportation and storage, large units would be more convenient.

Why do we Require Packaging?

Imagine there is no packaging material and technology available. How we are going to transport apple, mango etc. to distant markets? Definitely as loose. Then how can we weigh or count the fruits being transported or marketed? This would be very difficult. Had there been no packaging the commodity would have suffered great physical, chemical and microbiological losses. It would have been very difficult to handle, manage, transport and market the produce as loose and unpacked. Therefore, the packaging of fresh as well as processed commodities was needed to contain, unitize and protect the produce.

Functions of a Packaging Material

1. *Containment*: Package contains the product within it and prevents any leakage, pilferage and other losses.
2. *Unitization*: Package unitizes the product in suitable sizes for storage, transportation, marketing and distribution depending on the requirement.
3. *Protection*: Package protects the product from physical (by insects, animals, rodents, environmental i.e. rain, Sun etc.), chemical and microbiological spoilage.
4. *Identification and advertisement* : Packages of different brands of products may be of different shapes, colours and depict different attributes of the product. Package also displays important information about the contents. This is why we readily identify Frooti tetrapack or Pepsi bottle even from some distance and without reading anything.

Advantages of Packaging

- Prevention of physical damage
- Ease of management, handling, transportation, storage and marketing
- Reduction of qualitative and quantitative losses
- Package is a trademark and symbol of quality
- Package increases ease or convenience of use to the consumer in processed products.

Characteristics of a Good Packaging Material

1. Should protect the commodity from physical damages and hazards of handling and transportation.

2. Should protect the commodity from microbiological damages and chemical changes
3. Should be light weight, attractive, water resistant, temperature tolerant, printable
4. Should be non-toxic and non-reactive with the product
5. Should retain the quality and quantity of the produce
6. Should be transparent, low cost and temper proof
7. Should be environment friendly and biodegradable
8. Should not carry any unethical values
9. Should have desired barrier properties (i.e. permeability for individual gases and gas mixtures)
10. Closure of package should be heat sealable
11. Should be free from fogging due to respiration of fresh commodities packed inside.
12. Should filter out harmful UV light

Qualities of packaging material : A package should be able to:

1. *Protect* the produce from hazards of handling and transportation
2. *Prevent* microbial, insect and rodent damage
3. *Present* the contents in the best way
4. *Minimize* changes in quantity, physiological and biochemical quality.

Pre-Packaging

A very short definition of pre-packaging is "*to wrap or pack a commodity in small convenient units before marketing*" or "*to package (as food or a manufactured good) before offering for sale to the consumer*". The process of packaging of the produce in consumer size units either at producing centre before transport or at terminal markets is called as *pre-packaging*. The size of the unit may depend on the type of consumer. Generally we remove the inedible portion and pack only the edible portion eg cauliflower stalk may be removed and only the curd is packed. Pea may be packed after removing hull. Fruits and vegetables like beans carrots, radish, green chillies, brinjals, leafy vegetables, oranges, lemon, banana, grapes, and some flowers like chrysanthemum can be pre-packed.

The operation involves cleaning the fresh produce with water, trimming off unwanted and inedible portions, shade drying and packing in polyethylene bags (1/2 to 1 Kg) at the farm itself. The bags can be loosely tied at about 4 - 5 cm below from the top. The shelf life extends by 1.5–2 times under normal conditions, without any refrigeration and product remains fresh without any significant changes in nutritional and sensory quality. The pre-packaged produce has better consumer appeal and longer shelf-life besides having convenience in handling, use, transportation and marketing.

Advantages of Pre-Packaging

1. Reduces transportation cost by eliminating unwanted and inedible portion of fruits and vegetables
2. Requires less space for storage, shipping and transportation
3. Better presentation and eye appeal
4. Reduces shopping time as the produce is already graded
5. Easy to handle and better storage life

Limitations of Pre-Packaging

1. Adds to the costs
2. Not commonly practiced in India
3. May reduce storability of some commodities under ambient conditions.

Modified Atmosphere Packaging (MAP)

MA condition is generally characterized by the increased level of CO_2 and decreased levels of O_2 created by respiration of fresh fruits and vegetables or by active generation of MA as discussed in earlier chapters. When a commodity is packed in such a way that modified atmosphere i.e. reduction in O_2 and increase in CO_2 levels, is generated inside the package, it is called *modified atmosphere packaging*. In a modified atmosphere package, the product is exposed to the normal atmospheric gases (oxygen, nitrogen, carbon dioxide and water vapour) but the concentrations are different from those in the open air. The packaging consists of polymeric film pouch or plastic container with a specified gas permeability. Both fresh and processed products can be packed in MAP. The technology would be little different for fresh and processed products, but the ultimate aim is to reduce respiration rate and quality losses in fresh products and to reduce oxidative changes and loss of physical quality in processed products. Modified atmosphere can be created either passively by the product or intentionally by active packaging as discussed in previous chapters.

Examples of MAP in fresh commodities: Individual shrink wrapping of malta, capsicum, cucumber, brinjal, cabbage etc.

Examples of MAP in processed commodities: Uncle chips, Pepsi, Fanta, Vacuum packed tea.

Advantages of MAP Technology

- Increases shelf-life (50 to 400%)
- Improves presentation - clear view of the product and all round visibility.
- Hygienic pack having good stacking strength
- Product is sealed in package and is free from product drip, pilferage and odour.
- Cost reduction due to better utilization of labour, space and equipment.

Limitations of MAP Technology

- High capital cost of gas packaging machinery,
- High recurring costs of gases and packaging material
- Increased pack volume may increases transportation costs, storage and retail display space.

Gases Used in MAP

In MAP, the pack is flushed with a single gas (generally N_2 or CO_2) or a combination of gases i.e. oxygen, nitrogen and carbon dioxide. Traces of carbon monoxide, nitrous oxide, ozone, argon, ethanol vapour and sulphur dioxide are also used. Oxygen is kept at minimum levels because oxygen reacts with the foodstuff resulting in the oxidative breakdown of food and rancidity in fats and oils. Nitrogen is a good substitute to oxygen since it displaces oxygen from the pack and oxidative rancidity is delayed alongwith providing cushioning effect to products like potato or banana chips. Carbon dioxide has bacteriostatic and fungistatic effect in MA packaged food which is increased as the temperature is lowered because of increased solubility. It retards the growth of moulds and aerobic bacteria.

Materials Used for MAP

Flexible plastic packaging materials comprise of nearly 90% of the materials used in MAP due to their wide range of permeability to gases and water vapour, alongwith necessary package integrity needed for MAP while, paper, paperboard, aluminium foil, metal and glass containers account for only 10 % in MAP.

CONTAINERS FOR PACKAGING

Types of Containers/Packages

a. *Primary container / package* comes in direct contact with the food eg can, jar, trays, nets, bottles.

b. *Secondary container / package* is an outer box, wrapper, case that holds several unitized containers and prevent dirt and contaminants from spoiling the primary containers. eg. paperboard boxes, shipping cartons.

c. *Tertiary containers / package* group several secondary packages together. These are used for palletization at airports or transport terminals. eg *forklift shipping containers.*

A. Packaging Material used for Handling Fresh Fruits and Vegetables

1. *Plastic Crates / Boxes* : Used for on farm handling and transportation of fruits and vegetables. Now a days some special crates are provided by various companies for transportation of fruits from the field to the cold storage. eg. boxes provided by Adani Group for apple in Himachal Pradesh. These plastic crates are very sturdy, water resistant, ventilated and have good stacking strength.

2. *Wooden Boxes :* Earlier these boxes were used for the long distance transportation of fruits (apple, pear, mango, stone fruit etc.) and vegetables (tomato) but now they have become obsolete due to their various disadvantages i.e. environmental problems, injury to the fruits etc.

3. *CFB Cartons* : Corrugated Fibre Board Cartons are light in weight, easy to handle, recyclable, suitable for printing etc. These can also be made water resistant by wax coating or plastic coating. Available in 3 ply or 5 ply.

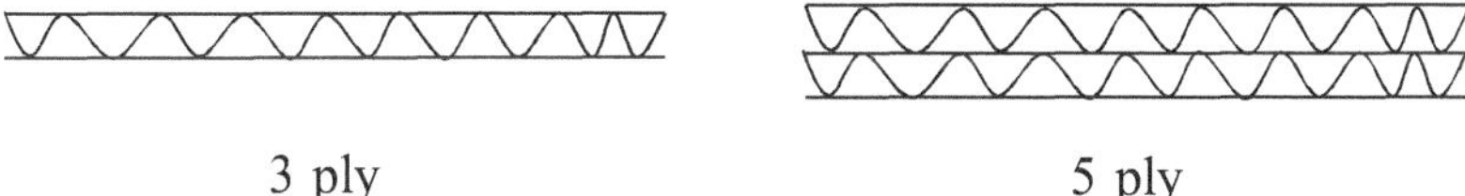

3 ply 5 ply

Advantages of CFB Boxes over the Conventional Wooden Boxes

a. Minimal bruising damage
b. Easy handling and stacking
c. More economical transportation
d. Can be punched, ventilated and printed at low cost
e. Made pilfer proof and reveal tampering at a glance
f. Occupy less space for storage of empty cartons
g. Can be easily made water proof

4. *Stretch Wrapping*: Used for retail marketing of fresh produce in the form of cling plastic films. They provide modified atmosphere and improve presentation of produce.
5. *Environment Friendly packaging*: This basically includes some biodegradable material used for packaging of fresh commodities. eg. packages made of sal (*Shorea robusta*) leaves and arecanut (*Areca catechu*) leaf sheath.

B. Packaging Material used for Handling Processed Products

1. Tin Cans

These are made of thin steel plate of low carbon content, thinly (about 0.000025 cm) coated on either side with tin metal.

Lacquering: It is difficult to coat steel plate uniformly with tin during the manufacturing process and some microscopic spaces are always left uncoated and open so the contents of CAN can react with metal and cause discolouration of product. Therefore, the inner side of the can is coated with lacquer which prevents discolouration of product and this process is known as *lacquering*.

Properties

a. Should not impart its own flavour to the product.
b. Should not injure the wholesomeness of the contents.

Two Types of Lacquers

a. *Acid resistant* : These lacquers are ordinary gold coloured enamel and cans when treated with it are called *R enamel cans*

   Uses: For packaging fruits of acid group with soluble colouring matter.

b. *Sulphur resistant* : These lacquers are also golden coloured and cans coated with it are called *C enamel cans*

   Uses : For packaging non-acid products like peas beans, corn, lima beans, red kidney beans etc.

2. Glass Containers

Glass containers are also one of the oldest containers used for packaging processed products. They possess some unique properties like non-reactivity with any type of food material packed inside, reusability, tolerance to heat, visibility/ presentation of the material. The only limitation in their use is their heavy weight and breakability. eg bottles, jars etc.

3. Flexible Packaging Material

These are the containers made of material which is not rigid like glass or tin containers but is flexible, therefore, is very suitable for packaging and transportation due to light weight, unbreakable nature and convenience of use (i.e. opening and reseal ability in some cases). Examples of the main flexible packaging materials are paper, plastic film and thin metal foil. Some films singly do not offer all the properties required for packing a particular product so many of these individual films are combined through processes like lamination and co-extrusion.

a. Polyolefins

a. *Low Density Polyethylene (LDPE)*

b. *Ethylene Vinyl Acetate*, a copolymer of ethylene and vinyl acetate having superior sealing qualities.

c. *Blended polyethylene* used to make a peelable seal, which is strong and gives an adequate barrier.

d. *Linear Low-Density Polyethylene (LLDPE):* Has better impact strength, tear resistance and tensile strength.

e. *High Density Polyethylene (HDPE):* Has a higher softening point than LDPE and provides superior barrier properties.

f. *Polypropylene (PP):* PP is chemically similar to polyethylene and can be extruded or co-extruded to provide as the sealant layer.

g. *Oriented Polypropylene (OPP):* Provides high moisture, gas and vapour barrier properties (7-10 times) than that of polyethylene.

h. *Co-extruded Oriented Polypropylene (COPP):* Used for vertical form-fill-seal packs.

i. *Inomers:* Polymer of ethylene, Surlyn A.

b. Vinyl Polymers

j. *Ethylene Vinyl Acetate Co-polymer (EVA)*

k. *Poly Vinyl Chloride (PVC):* Has excellent oil and grease resistance. Barrier properties and strength characteristics vary with thickness.

l. *Polyvinylidene Chloride (PVdC) Co-polymer*

m. *Ethylene Vinyl Alcohol (EVOH)*

c. Styrene Polymers

n. *Polystyrene*

o. *High-impact Polystyrene*

d. Polyamides

p. Nylon-6

e. Polyesters

q. Polyethylene Terephthalate (PET)

f. Other Polymers

r. polycarbonate

s. Acrylonitrite Butadieine Styrene (ABS)

Specific Packaging Technology

1. *Hermetic closures* : *Hermetic* refers to the container that is sealed completely against the entry of gasses and vapours. These are also impervious to bacteria, yeasts, molds and dirt. Hermetic sealing is essential for vacuum and gas packages. The containers preventing the entry of microorganisms only may not be impervious to gases and vapours since the size of gas particles is smaller than microorganisms eg. cans, tetrapacks.

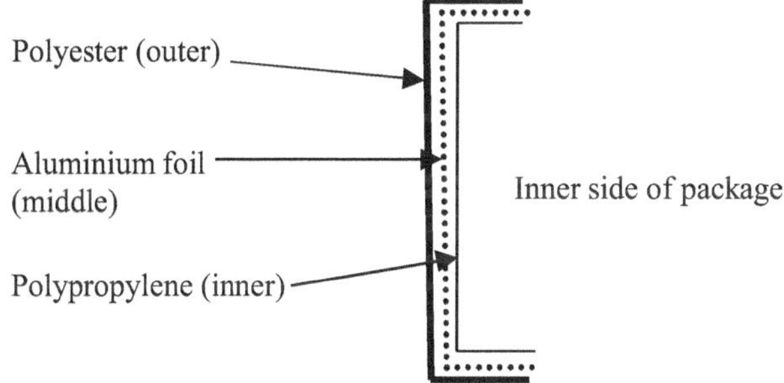

Fig. Retortable Pouches

2. *Form Fill and Seal packages*: The package (generally pouches) of different capacities are preformed from flat rolls of the packaging material (*Form*) then filled with a fixed quantity of food product (*Fill*) and then is sealed air tight (*Seal*). eg Kurkure pack, pan masala packs etc.

3. *Retortable pouches, trays*: These are suitable for packing low acid foods which need to be retorted or heated at high temperatures. These generally consist of three layers

(i) *Polyester* : high temperature resistance, strength, printability

(ii) *Aluminium foil*: barrier properties

(iii) *Polypropylene*: heat seal ability

4. *Stretch films and shrink films*: The shrink films are used for individual as well as group packaging of fresh as well as processed products. These films have the ability to shrink when cooled after passing through hot air and hold the product in its position.

5. *Aerosol cans*: Foods generally semi-solid are packed in sealed cans under pressure (8-18 atmosphere) with inert gas or sterile air mixture (used as propellant). When the nozzle is pressed (as in case of perfume packs) the food comes out generally in the form of foam.

6. *Coextruded films*: These films are formed by coextrusion in 3, 5 or upto 9 layers with material i.e. Nylon, HDPE, EVA, Metallocene, LLDPE (all types), VLDPE, LDPE.

7. *Tetrapacks* : These are the packages used extensively for packing various types of products i.e. Fruit juices, Verka milk, Vegetable Oils (Dhara) etc.. These packages have 6-7 layers of materials used for their manufacture

   1. *Polyethylene*: Innermost layer suitable for sealing
   2. *Polyethylene*: Adhesion layer
   3. *Aluminium* foil: Gas and light barrier, aroma retention
   4. *Polyethylene*: Adhesion layer
   5. *Paper*: Stability, strength and printing
   6. *Polyethylene*: Outer most layer makes the package water/ moisture proof

8. *Reseal able pouches*: Some foods now come in a flexible pouch package provided with a zipper for sealing the pack again and again after use eg. Haldiram's Namkeen and Snacks.

9. *Laminates*: Multilayered packages made form paper, plastic films and metal foils having low permeability to moisture and light, better retention of quality, heat seal ability etc.

Points to Remember

√ Packaging of fruits is chiefly done to assemble the produce in convenient units for storage, marketing and distribution.

√ Containment, unitization protection and advertisement are the main functions of packaging

√ The process of packaging of the produce in consumer size units either at producing centre before transport or at terminal markets is called as pre-packaging.

√ When a commodity is packed in such a way that modified atmosphere i.e. reduction in O_2 and increase in CO_2 levels is generated inside the package, it is called modified atmosphere packaging.

√ Examples of MAP in fresh commodities: Individual shrink wrapping of malta, capsicum, cucumber, brinjal, cabbage etc.

√ Examples of MAP in processed commodities: Uncle chips, Pepsi, Fanta, Vacuum packed tea

√ Oxygen, nitrogen and carbon dioxide along with traces of carbon monoxide, nitrous oxide, ozone, argon, ethanol vapour and sulphur dioxide are the gases used un MAP.

√ Flexible plastic packaging materials comprise of nearly 90% of the materials used in MAP due to their wide range of permeability to gases and water vapour, along with necessary package integrity needed for MAP while, paper, paperboard, aluminium foil, metal and glass containers account for only 10 % in MAP.

√ *Primary container/package* comes in direct contact with the food eg can, jar, trays, nets, bottles.

- √ *Secondary container / package* is an outer box, wrapper, case that holds several unitized containers and prevent dirt and contaminants from spoiling the primary containers. eg. paperboard boxes, shipping cartons.
- √ *Tertiary containers / package* group several secondary packages together. These are used for palletization at airports or transport terminals. eg *forklift shipping containers*
- √ The inner side of the can is coated with lacquer which prevents discolouration of product and this process is known as lacquering.
- √ Acid resistant lacquers are ordinary gold coloured enamel and cans when treated with it are called R enamel cans and used for packaging fruits of acid group with soluble colouring matter.
- √ Sulphur resistant lacquers are also golden coloured and cans coated with it are called C enamel cans and are suitable for packaging non-acid products like peas beans, corn, lima beans, red kidney beans etc.
- √ CFB stands for Corrugated Fibre Board and the cartons comes in 3 ply or 5 ply
- √ Packages made of sal (*Shorea robusta*) leaves and arecanut (*Areca catechu*) leaf sheath are eco-friendly.
- √ Retortable pouches are made of three layers i.e. polyester, aluminium foil and polypropylene
- √ Tetra packs are made of 6 layers of material i.e. Polyethylene, Polyethylene, Aluminium foil, Polyethylene, Paper, Polyethylene.

6

General Principles and Methods of Food Preservation

Nature has made such a beautiful system that a fruit does not spoil unless its outer covering is removed and till it is there on the plant. For example banana will not spoil for long time if we do not remove its peel. Similar is the case with apple, guava, mango and many other fruits and vegetables. But as soon as the outer covering is removed the damage begins and the commodity is spoilt very soon.

Preservation is extending shelf life of fresh or processed food by using various methods and prevention or reduction of spoilage. Better the preservation, lesser the spoilage, more the spoilage poorer and inefficient is the preservation.

Therefore, in order to prevent or reduce the spoilage, we have to first identify and understand the causes of spoilage. There are chiefly 3 agencies that cause spoilage in fruits and vegetables i.e. (i) microorganisms, (ii) enzymes & chemical changes and (iii) man, animals, rodents, insects, etc. For effective preservation we have to check the growth and activity of these spoilage causing agencies. So, in order to prevent spoilage and preserve a perishable commodity for long time the following principles and methods are to be followed.

Principles of Food Preservation

S.No.	Principle	Method
1	*(Prevention or delay of microbial decomposition)*	
	a by keeping out the microorganisms	Aspesis, packing
	b by removal of microorganisms	Washing, trimming, filtration, centrifugation etc
	c by hindering the growth and activity of microorganisms	Low temp., drying, anaerobic conditions, chemicals, sugar, salt, acids, oil & spices etc.
	d by killing of microorganisms	Heat, radiation, high pressures
2.	*(Prevention or delay of self decomposition of the food)*	
	a by destruction or inactivation of food enzymes	Blanching
	b by prevention or delay of purely chemical reactions	Use of antioxidants, sulphites
3.	*(Prevention of damage because of insects, animals, rodents, mechanical causes etc.)*	Packing, mechanical means etc.

Among all the above three agencies, microorganisms is the one which is very important and is considered to be most potential for causing spoilage and damage to the quality. If not addressed properly, the severity may be such that eating the spoilt food may also lead to death of the consumer in some cases. Enzymes and chemical reactions may not be very potent and merely lead to reduction in sensory and nutritional quality attributes. The third agency i.e. insects, animals, rodents etc. may cause only physical loss to the commodity and can be easily controlled by proper packing and handling.

But there is definite correlation between the damages caused by all the three agencies eg. if an apple is eaten partially by rat, it may open the fruit tissues to air and allow the entry of microorganisms as well as oxygen leading to spoilage caused by microorganisms or self decomposition (browning of tissues) by enzymatic reactions.

1. Prevention or Delay of Microbial Decomposition

a. Keeping out the Microorganisms (asepsis)

As discussed earlier, most of the fruits, i.e. apple, banana, mango will not spoil for long time if their outer covering is not removed. Reason for this would be that the spoilage causing microorganisms from the air cannot enter the fruit tissues through the fruit skin and thus prevention of spoilage for long time. Initial researchers also got an idea why not to do something by which the entry of micro-organisms into the fruit tissues can be checked. This lead to the development of first section of the first principle i.e. keeping out the microorganisms. This can be achieved by aspesis i.e. not allowing the entry of microorganisms into the fruit by not opening the natural fruit covering/skin or by proper packaging which would not allow the entry of microorganisms.

b. Removal of Microorganisms

If somehow some microorganisms have entered the fruit, the intensity of spoilage caused by them can be reduced by removal of some of them i.e. reducing microbial load. This can be achieved by filtration through membranes and centrifugation (in juices), trimming off the damaged or spoilt tissues (cabbage, cauliflower, apple, banana etc) or washing the fruit for removal of adhering microorganisms (tomato, apple, mango etc).

c. Hindering Growth and Activity of Microorganisms

The above method of removal of microorganisms may not remove all the microorganisms from the food, therefore, we should also provide unfavourable conditions to the microorganisms so that their growth, activity and multiplication is checked. These unfavourable conditions may be low temperature (freezing), anaerobic conditions (carbonation, use of KMS), osmotic or hypertonic solutions (high sugar and salt concentrations i.e. jam, jellies, honey, pickles etc.), low pH (vinegar, pickles, lime and lemon juice), low water activity (drying & dehydration, intermediate moisture foods), high alcohol (fermented beverages), presence of chemicals changing microbial structure and function (preservatives etc.).

d. Killing of Microorganisms

The final solution in food preservation would be killing of microorganisms which can be achieved by applying heat of variable intensities (pasteurization, sterilization), radiation (also called as cold sterilization), high pressures (physical rupture of microbial cells at low temperature). But the microorganism may enter the commodity if it is not packed properly (aseptically) after killing of microorganisms.

It is observed that a single method would not be sufficient for preservation of a commodity. More than one methods should be combined for effective preservation.

2. Prevention or Delay of Self Decomposition

a. Destruction or Inactivation of Food Enzymes

You might have observed that potato or apple when cut turns brown over the cut surfaces on keeping open for some time. Whereas boiled potatoes when cut do not turn brown. This is because of the reason that application of heat during boiling of potato inactivates the enzymes (PPO, Polyphenol oxidase) responsible for the said browning of potato or apple.

So this principle of preservation deals with the inactivation of enzymes for delaying or reducing spoilage. This can be achieved by blanching of the material which aims at inactivation of natural enzymes present in the food.

b. Prevention or Delay of Purely Chemical Reactions

You must also have observed that fruit juices if kept for long time turn brown, or fats (fatty foods) turn rancid on prolonged storage. These changes occur due to the purely chemical reactions going on within the food products. Juices turn brown due to non-enzymatic (Maillard) reactions and fats turn rancid due to their oxidation.

So this principle of preservation deals with reduction/ avoidance of spoilage caused due to purely chemical reactions. This can be achieved to some extent by proper packaging and use of anti-oxidants.

3. Prevention of Damage because of Insects, Animals, Rodents, Mechanical causes etc.

This principle of preservation deals with prevention of damage caused by various external agencies other than microorganisms and enzymes i.e. animals, man, insects, rodents etc. These agencies generally cause physical damage to the food material eg. Rats may eat peels of oranges in a storage, animals may also eat the food if kept within their reach etc. But in none of the cases these damages are deleterious to human health. If you consume a half eaten apple or orange you are generally never going to die or experience any health risks, but if the food is spoilt by microorganisms, and you consume the spoilt food your health shall definitely be at risk. The damaged food by animals, man, insects, rodents etc. may later on give way for the initiation of microbial and self decomposition.

Proper packaging of the food is predominantly the effective solution for prevention of damage caused by the agencies considered under this principle of preservation.

Overall, from food processors point of view the three principles are to be considered in decreasing order of importance and emphasis. Highest emphasis is given on control of microbial decomposition followed by self decomposition, ultimately followed by damage caused by animals, insects, rodents etc.

Methods of Food Preservation

a. Asepsis

Nature has made such a beautiful system that a number of commodities are naturally covered with some covering thus exemplifying asepsis. Apple and banana does not spoil for long till their peel is safe and intact. Similarly milk does not generally spoil till it is inside the body of cow. But as soon as the covering is removed and the tissues are exposed to the atmosphere spoilage begins. So asepsis is all about how we can keep the microorganisms out and away from the fruit edible parts.

This is now being emphasized upon to a great extent in food processing and preservation. Lesser the contamination better would be the preservation. *Aseptic Processing* and *Packaging* (APP) is also based on packing and sealing a microorganism free product in a microorganism impermeable package inside a microorganism free environment.

b. Removal of Microorganisms

If somehow the microorganisms have entered into the food or have settled over the surface of food, the microbial load i.e. the number of microorganisms should be reduced. This can be achieved in the following ways

- By washing — fruits and vegetables like apple, mango, tomato, radish, carrot etc.
- By trimming — potato, root vegetables, cabbage etc.
- By centrifugation — fruit juices and drinking water
- By sedimentation — fruit juices, drinking water etc.
- By filtration — fruit juices, drinking water, beverages, soft drinks, wine, beer etc.

c. High Temperatures

The microbial load can also be reduced by killing them by application of heat in variable intensities. Application of heat also inactivates the enzymes responsible for causing spoilage of various types in the food.

Thermal death time : Time taken for killing stated number of microorganisms or their spores at a certain temperature under defined conditions.

Thermal death point : Temperature at which all microorganisms present in the food are killed in 10 minutes.

Decimal reduction time : Time required to cause 90% reduction in the viable microbial cells/spores at a certain temperature.

Pasteurization : Heat treatment at temperatures below 100°C that kills a part but not all of the microorganisms present in the food.

Sterilization : Heat treatment generally above 100°C aimed at killing of all the microorganisms.

Simmering : Gentle boiling at about 100° C.

Blanching : Heat treatment in boiling water at about 100°C for 3-5 min. given to inactivate enzymes present in the food followed by immediate cooling to room temperature.

Frying : Heat treatment where outer food temperature is very high generally above 100°C but internal temperature does not reach 100°C.

Roasting : Internal food temperature reaches about 60-85°C in meat.

Cooking : Thermal processing for a specific time at a specific temperature. It is a very broad and indefinite term.

Canning : Process of preservation of various foodstuffs including fruits and vegetables whole or in pieces, in sugar syrup or brine by heat processing them in hermetically sealed containers.

Difference between Pasteurization, Sterilization and Blanching

Pasteurization	Sterilization	Blanching
1. Aims at killing of most but not all of the microorganisms	1. Aims at killing of all of the microorganisms	1. Aims at inactivation of enzymes rather than killing of microorganisms
2. Temp. < 100° C	2. Temp. > 100° C	2. Temp. = 100° C (boiling water)
3. Generally used in fruits	3. Generally used in vegetables	3. Used in both fruits and vegetables (but only solid pieces). We cannot blanch liquids.

d. Low Temperatures

Low temperatures are undesirable for the proper growth and multiplication of microorganisms therefore it hinders the growth and activity of microorganisms.

Freezing : Cooling the food so that the moisture inside the food converts into ice and is not available for microorganisms for their growth and activities which are definitely reduced at such low temperatures.

Sharp (Slow) freezing : Freezing in air at -15 to -29°C temperature generally within 3-72 hrs.

Quick freezing : Freezing the food at -17.8 to - 45.6 °C in less than 30 min.

Dehydrofreezing : Dehydration + freezing. Half of the moisture is first removed by dehydration and then the food is frozen.

e. Drying and Dehydration

Drying and dehydration involves removal of moisture from the food, creating low water activity which hinders the growth and activity of microorganisms and food is preserved for long time. It is infact the oldest methods of food preservation. It may be done by

- Keeping the food in open Sun
- Application of high temperatures (50-60°C) in air
- *Freeze drying* (lyophilization), direct sublimation of ice to vapours followed by their removal from the food
- Keeping the food in high concentration sugar solution for 24-48 hrs followed by final drying in hot air (*osmotic dehydration*)

f. Use of Additives (preservatives)

The addition of preservatives is based on the fact that these chemicals react with the food constituents or the microorganisms and create unfavourable conditions and hinder the growth and activity of microorganisms.

- *Class I preservatives* : sugar, salt, oil (do have any maximum limits of their use).
- *Class II preservatives* : Potassium metabisulphite (KMS), Sodium benzoate, Sorbic acid etc. (do have maximum permissible limits of their use as per law).

g. Anaerobic Conditions

Absence of oxygen also hinders the growth and activity of microorganisms. Anaerobic conditions can be generated by creation of vacuum, addition of CO_2 (carbonated beverages Pepsi, Coca cola, Fanta, Dew etc.) and addition of inert gas i.e. N_2 (Potato Chips, Banana Chips). This not only retards microbial growth and activity but also prevents oxidative changes in the nutritional composition of food.

h. Irradiation

Killing of microorganisms by exposure of food to various ionizing radiations generally produced by Cobalt - 60 and Cesium 137 under controlled conditions without generation of heat. Irradiation is also called "*Cold sterilization*".

i. Modern Methods

Besides above methods some new methods are also used for hindering the growth and activity and killing of microorganisms which shall be discussed in later part of the book. These methods are

- High pressure processing
- Use of pulsed electric and magnetic fields
- Use of pulsed light treatment
- Ohmic heating
- Ultra sounds
- Linear induction electron accelerator

Classification of Preservation Methods

Short term/Temporary preservation methods	Long term/Permanent preservation methods
Asepsis, anaerobic conditions, low temperatures, pasteurization, filtration, centrifugation, blanching, sedimentation, trimming, washing etc.	Sterilization, canning, drying, fermentation, high pressure processing, aseptic processing and packaging, irradiation, use of preservatives etc.

Classification of Food Products based on Shelf Life

1. *Non-perishable* : Food that can be stored under room conditions for months together. These contain very low moisture content generally < 8-10% eg. sugar, flour, dry beans

2. *Semi-perishable* : Food that can be stored under room conditions from few days to about a month or two eg. potato, nuts, almond

3. *Perishable* : Food that can not be kept in safe and sound condition at room temperature for more than a day or two. These contain very high concentrations of moisture generally > 70-80% eg milk, meat, fruits, leafy vegetables

pH Concept

pH is the -ve logarithm of hydrogen ion concentration in a product or logarithm of the number of litres of solution which contains 1 g of hydrogen ions

If 500 litres of a solution (a) contain 1g hydrogen ion, then pH will be

$$\text{Log } (500) = 2.6989$$

If 20,000 litres of solution (b) contain 1g hydrogen ions, then pH will be

$$\text{Log } (20{,}000) = 4.3010$$

If 1,00,000,000 litres of solution (c) contain 1g hydrogen ions, then pH will be

$$\text{Log } (1{,}00{,}000{,}000) = 8.0000$$

The concentration of hydrogen ions in solution (a) is maximum and in solution (c) the minimum therefore, the pH is higher in solution (c) i.e. towards alkaline and lower in solution (a) i.e. acidic.

Classification of Food Products based on pH (Cameron 1940)

1. *Low Acid Foods* : pH > 5.0 eg peas, beans, corn, meat, fish, poultry milk
2. *Medium Acid Foods* : pH 5.0-4.5 eg. meat, vegetables, soups, spinach, asparagus, pumpkin
3. *Acid Foods* : pH 4.5-3.7 eg tomato, pears, figs, pineapple
4. *High Acid Foods* : pH <3.7 eg. pickles, citrus juice, aonla, berries, sauerkraut

pH 4.5 is regarded as the dividing line between acid and non acid foods and pH 7.0 between acidic and alkaline

Note : During heat processing fruits are heated in boiling water at 100° C while the vegetables (except tomato) at 115-121.1° C under pressure because most of the fruits are acidic in nature having low pH. As a thumb rule bacteria does not prefer to grow at low pH while molds and fungus do. On the contrary low acid foods are first spoilt by bacteria rather than molds and fungus. This is the reason that we seldom find fungus growing on milk but often find fungus growing on spoilt citrus fruits.

$$100° C = 212 °F$$

For every rise in altitude of 500 ft or 152 m the boiling point of water reduces by about 1°C

$101° C \cong 1 lb/in^2$

$115° C \cong 10 lb/in^2$

$121.1° C \cong 15 lb/in$

Points to Remember

- √ *Preservation* is extending shelf life of fresh or processed food by using various methods and prevention or reduction of their spoilage. Better the preservation, lesser the spoilage, more the spoilage poorer and inefficient is the preservation.
- √ 3 agencies that cause spoilage in fruits and vegetables are (i) microorganisms, (ii) enzymes & chemical changes and (iii) man, animals, rodents, insects, etc.
- √ The 3 principles of preservation deal with prevention or delay of decomposition or damage caused by the above 3 agencies causing spoilage.
- √ By low temp., drying, anaerobic conditions, chemicals, sugar, salt, acids, oil & spices etc. we can only hinder the growth and activity of microorganisms while application of heat, irradiation, high pressures can kill the microorganisms.
- √ Keeping out the microorganisms is called as asepsis.

- √ Microorganisms can be removed by washing, trimming, centrifugation, sedimentation, filtration etc.
- √ Boiled potatoes when cut do not turn brown due to inactivation of PPO enzyme during boiling.
- √ *Pasteurization is the* heat treatment at temperatures below 100°C that kills a part but not all of the microorganisms present in the food. While, *Sterilization* is the heat treatment generally above 100°C aimed at killing of all the microorganisms and *Blanching* is heat treatment in boiling water at about 100°C given to inactivate enzymes present in the food for about 3-5 min followed by immediate cooling to room temperature.
- √ *Canning* is preservation of food whole or inpieces in sugar syrup or brine by heat processing in hermetically sealed containers.
- √ *Sharp (Slow) freezing* is the freezing done in air at -15 to -29°C temperature generally within 3-72 hrs while, *Quick freezing* is completed at -17.8 to - 45.6 °C in less than 30 min.
- √ Drying and dehydration involves removal of moisture from the food, creating low water activity which hinders the growth and activity of microorganisms and food is preserved from long. It is infact the oldest methods of food preservation.
- √ Sugar, salt, oil which do not have any maximum limits of their use are grouped as Class I preservatives while, Class II preservatives do have maximum permissible limits of their use and examples are Potassium metabisulphite (KMS), Sodium benzoate, Sorbic acid etc.
- √ Anaerobic conditions can be generated by creation of vacuum, addition of CO_2 (carbonated beverages Pepsi, Coca cola, Fanta, Dew etc.) and addition of inert gas i.e. N_2 (Potato Chips, Banana Chips). This not only retards microbial growth and activity but also prevents oxidative changes in the nutritional composition of food.
- √ Freeze drying is called as lyophilization and irradiation is called as Cold sterilization.
- √ pH is the -ve logarithm of hydrogen ion concentration in a product or logarithm of the number of litres of solution which contains 1 g of hydrogen ions.

- √ pH 4.5 is regarded as the dividing line between acid and non-acid foods
- √ Bacteria do not prefer to grow at low pH while molds and fungus do. On the contrary low acid foods are first spoilt by bacteria rather than molds and fungus.
- √ 100° C = 212 °F; 101° C (1 lb/ in ^{2}); 115° C (10 lb/ in ^{2}) ; 121.1° C (15 lb/ in^2).

7

Tomato Products

Essentials for Quality Product Preparation

- Plant ripened red tomatoes should be selected. Yellow and green fruits on processing yield dull colour and turn brown due to oxidation.
- Never use iron equipment for processing. Lycopene oxidizes and turns brown when comes in contact with iron. It also forms black compounds with the tannins present in tomatoes or the spices which are used for product preparation.
- Avoid prolonged heating and cool immediately to avoid over-cooking.

Product Specifications

Product	Min TSS (° Brix)	Remarks
Juice	5	Not more than 0.5% (w/w)
Soup	7	salt, sugar, dextrose, malic acid, ascorbic acid, citric acid and permitted colours (for juice and soup)
Puree		
Medium	9	
Heavy	12	
Paste	25	
Ketchup	25	1.0 acidity
Sauces mixed	15	1.2 acidity

Characteristics of Juice (raw material)

- Should be deep red in colour
- Should possess the characteristic taste and flavour of tomato
- Acidity should be about 0.4% (as citric acid)
- Should be uniform in quality

Crushing : Fruits are cut into 4-6 pieces and crushed in a fruit mill.

Pulping : 2 methods

A. Hot Pulping / Hot Process

Crushed tomatoes are boiled in their own juice in stainless steel pans for 3-5 min.

Advantages

1. The tendency of juice to separate into liquid and pulp can be overcome due to release of pectin into the juice and inactivation of pectase enzymes

2. Partial sterilization of juice is achieved due to reduction in microbial load.
3. Red colour is also released during cooking
4. Higher yields.

B. Cold Pulping / Cold Process

Tomatoes are crushed and passed through pulper without boiling.

Disadvantages

1. Less yield
2. Incorporation of air into juice / pulp, so, encourages oxidation
3. Less red colour released into juice
4. Chances of microbial spoilage are more.
5. Separation of juice into watery portion and pulp.

1. Tomato Juice

Juice Extraction : It is done through different types of juice extractors, namely Continuous Spiral Press and Cyclone or Pulper.

Total Solids : Generally the total solids are 5.66% (FPO not less than 5%). Specific gravity 1.0240 at 20° C. About 0.5 % salt and about 1% sugar can also be added.

Packaging : Done in glass bottles or cans. Before packaging, the juice is generally homogenized to retard separation of liquid from the pulp. Juice is heated at 88° C and filled hot into glass bottles followed by hermetically sealing and sterilization in boiling water (100°C) for 30 min.

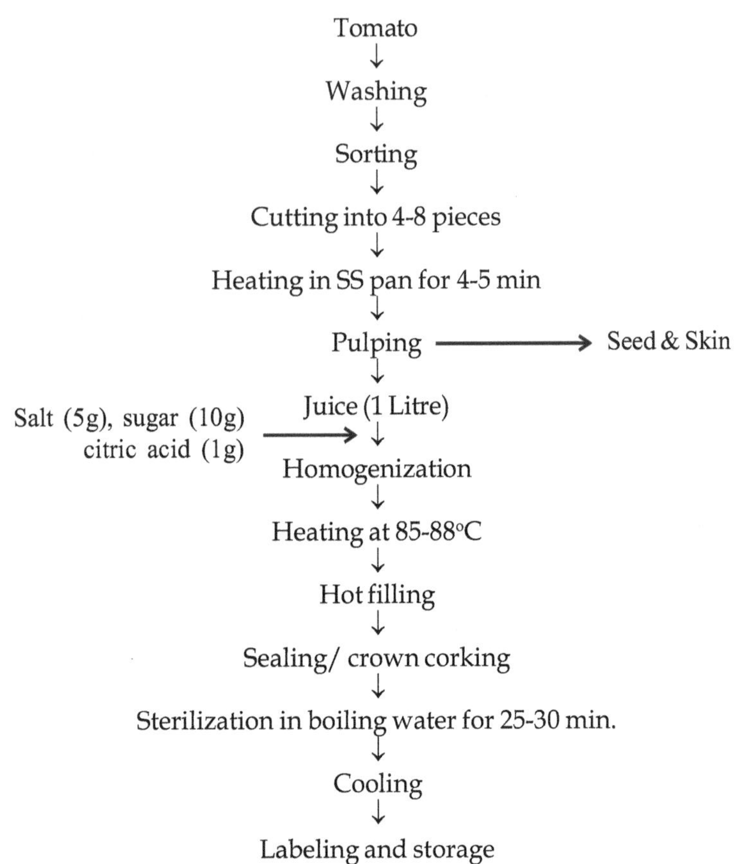

Preparation of Tomato Juice

2. Tomato Puree and Paste

Tomato pulp devoid of or free from skin and seeds, with or without added salt and containing not less than 8.37% salt free tomato solids is called as called as *"medium tomato puree"*. It is further concentrated to 12% solids to form *"Heavy Tomato Puree"*. Pulp is extracted and concentrated in open cookers or vacuum pans and packaging is done in glass bottles or cans.

Tomato Paste is a concentrated tomato juice or pulp free from skin and seeds, containing not less than 25% of tomato solids. *Concentrated tomato paste* contains 33% tomato solids.

Common salt, basil leaf or sweet oil of basil leaves may also be added.

Tomato paste is used as a semi-finished raw material for the final conversion into ketchup by many organizations i.e. Nestle etc. Tomato are available in plenty during May-June in North Indian plains and the crop is damaged as soon as monsoon comes during June end-July beginning. Therefore, tomato fruits are converted into paste and preserved using aseptic packaging and processing in big aluminium laminated bags of about 200 Kg capacity, stored and transported to different places for conversion into ketchup and other products during lean season.

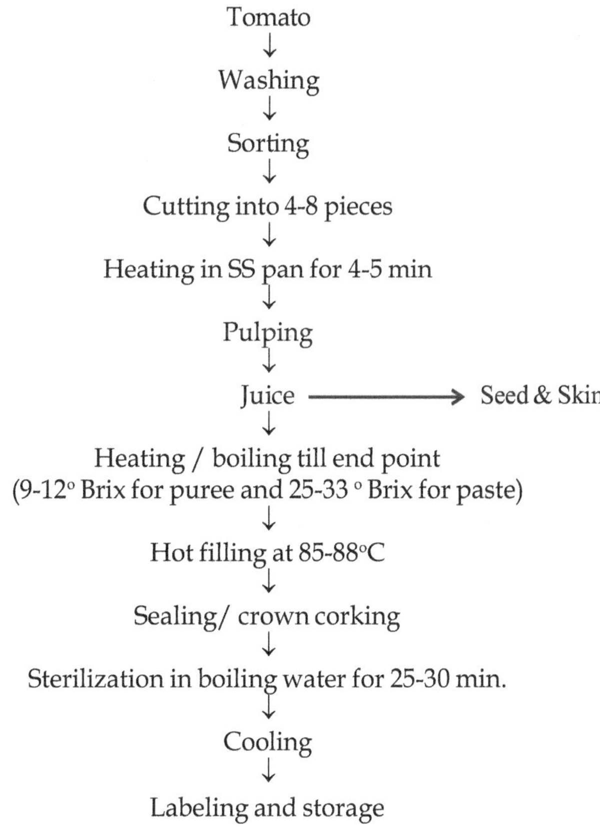

Preparation of Tomato Puree and Paste

3. Tomato Cocktail

Product prepared by adding spices including chilli, cloves, cardamom, coriander alongwith vinegar and common salt to the tomato juice. Lemon or lime juice etc may also be added in different proportions to suit the palate.

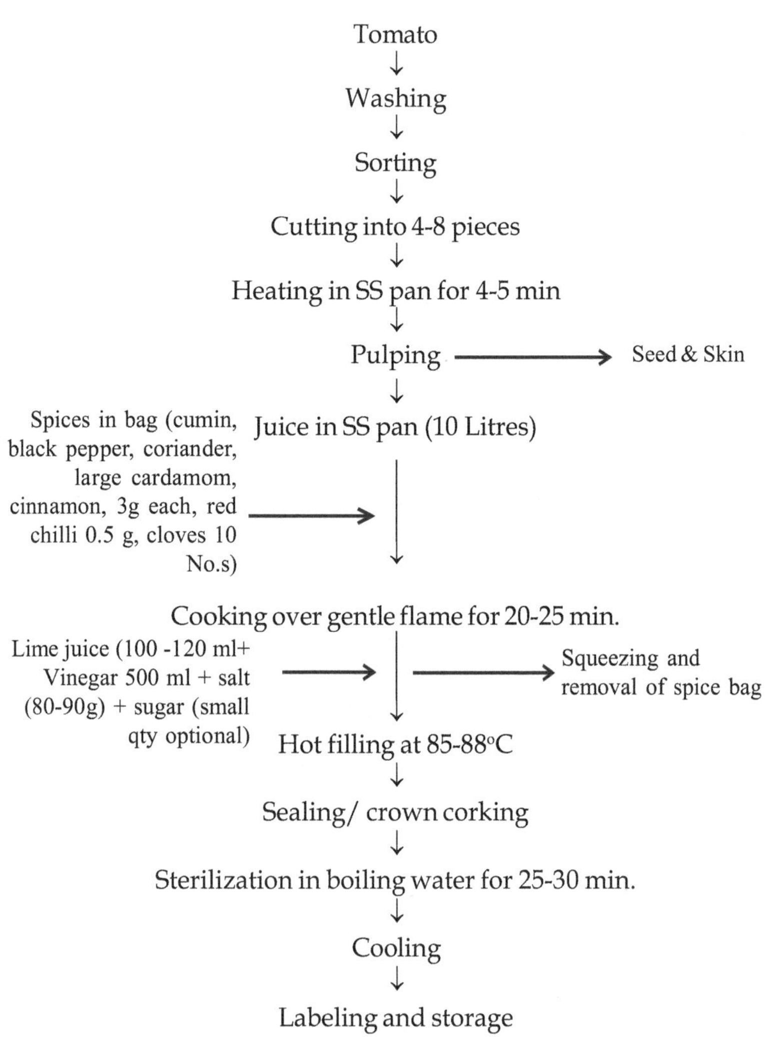

Preparation of Tomato Cocktail

4. Tomato Ketchup and Sauce

Ketchup is a product made by concentrating tomato juice or pulp without seeds and skin, with added spices, salt, vinegar, onion, garlic etc. so that it contains not less than 12 % tomato solids and generally 28% or more total solids (not less than 25% TSS as per FPO specifications).

Sauce is a product similar to ketcup but may be prepared from pulp of tomato or other fruits/vegetables and have lesser TSS (not less than 15%) and is thinner in consistency.

Addition of Ingredients

(i) *Spices* : Good quality spices should be added in proper proportions by the following methods.

a. *Bag method* : Spices/powders are tied loosely in a muslin cloth bag and placed in boiling tomato juice.

b. *Spice extracts* : Spices are boiled for long time to prepare extract which is added to the juice during ketchup preparations.

(ii) *Sugar* : 1/3rd of sugar is added initially in order to fix the red colour in tomato. Rest of the sugar is added a little before the ketchup is ready. Sugar is added @ 6-8 % of the tomato juice depending upon the recipe (10-26%) in finished product.

(iii) *Salt* : Salt is added towards the end of boiling @ 0.8% to 1.0% so that the finished product contains 1.3-3.4% salt.

(iv) *Vinegar* : Ketchups contain 1.25-1.50 % acetic acid. Vinegar should also be added towards the end when the ketchup has thickened sufficiently. Otherwise the acetic acid may get volatized leaving the ketchup deficient in acid as well as flavour.

(v) *Thickening agents* : Pectin is usually added @ 0.1-0.2% by weight of the finished product.

(vi) *Preservatives* : Sodium Benzoate @ 750 ppm may be added in finished product.

Concentration : Commercial ketchups have generally 28-30% total solids and may be as high as 37%. The increase in solids increases its keeping quality.

Bottling : Ketchup is filled in bottles at 88° C and are pasteurized for 30-35 min. in hot water at 85-88° C after corking. It is preferable to add 250ppm sodium benzoate and then pasteurize the product.

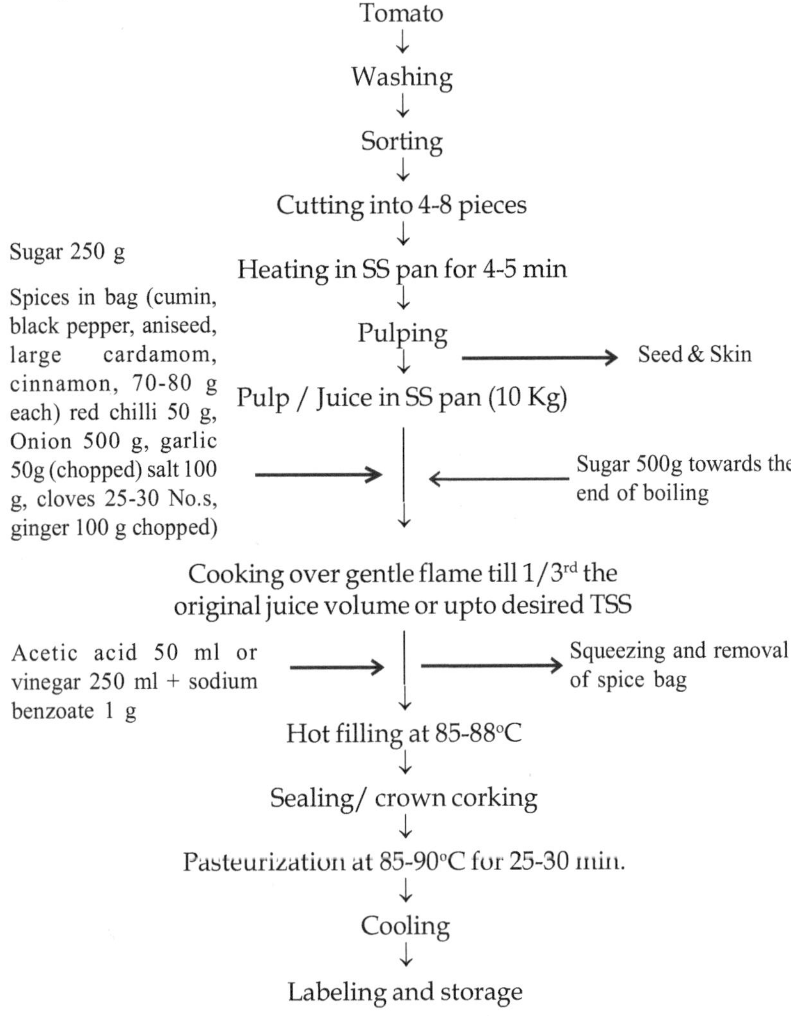

Preparation of Tomato Ketchup and Sauce

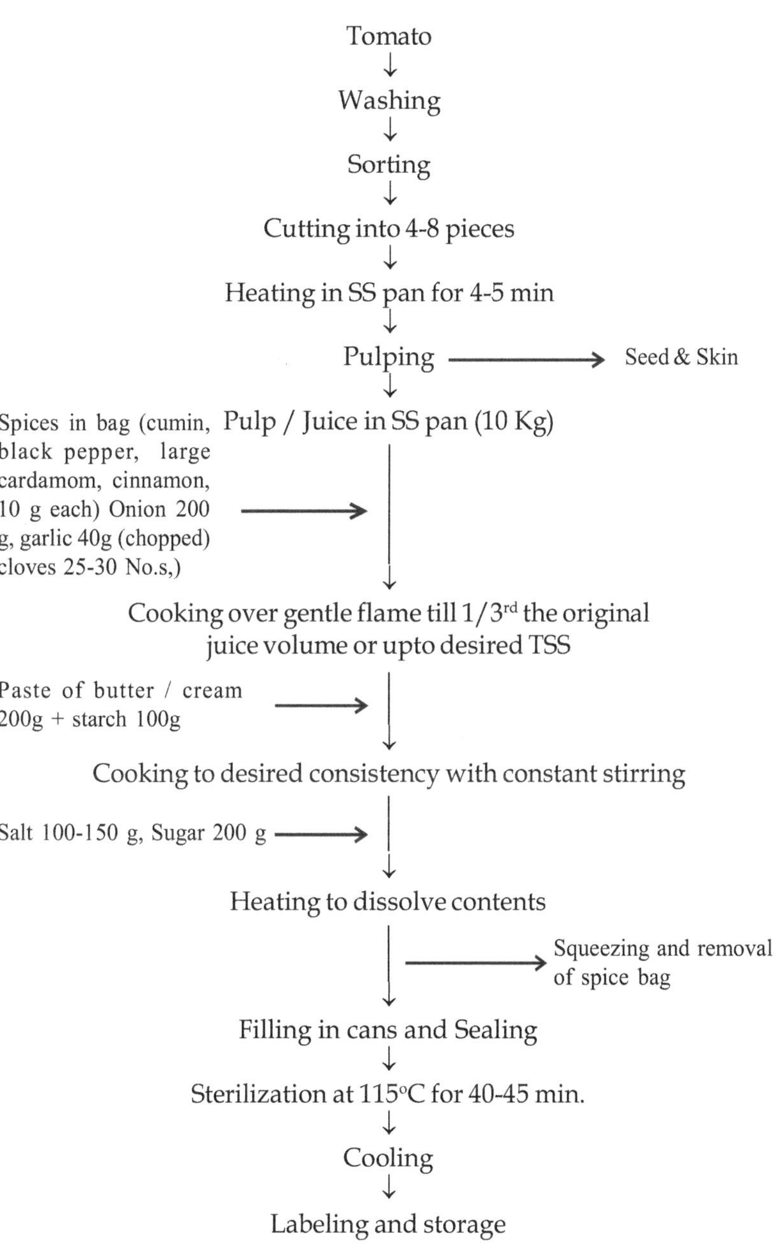

Preparation of Tomato Soup

Difference between Ketchup and Sauce

Ketchup	Sauce
1. Prepared from tomato only	1. Prepared from tomato as well as other fruits such as pumpkin, chilli etc.
2. Min. TSS 25%	2. Min. TSS 15%
3. Min. acidity 1.0%	3. Min. acidity 1.2%
4. Thicker in consistency	4. Thinner in consistency
5. Costly	5. Cheap as compared to ketchup
6. Only red in colour	6. May have, red, green or other colours

5. Tomato Soup

Tomato soup is a product having TSS not less than 7 °Brix as per FPO, prepared from tomato juice and which is served hot and is relished equally before principal meals of the day as well as during any other time.

Points to Remember

- √ Plant ripened red tomatoes should be selected for preparing good quality product. Yellow and green fruits on processing yield dull colour and turn brown due to oxidation.
- √ Lycopene oxidizes and turns brown when comes in contact with iron. It also forms black compounds with the tannins present in tomatoes or the spices which are used for product preparation.
- √ Tomato juice should be deep red in colour, having characteristic taste and flavour of tomato
- √ In Hot Pulping / Hot Process the crushed tomatoes are boiled in their own juice in stainless steel pans for 3-5 min and then passed through pulper. This overcomes the tendency of juice to separate into liquid and pulp due to inactivation of pectase enzymes and releases more red colour in the juice.

- √ In Cold Pulping / Cold Process the tomatoes are crushed and passed through pulper without boiling resulting in lesser yield and lesser red colour released into juice.
- √ Salt, sugar and acid may be added in tomato juice.
- √ Tomato pulp free from skin and seeds, with or without added salt and containing not less than 8.37% salt free tomato solids is called as *medium tomato puree* and one having not less than 12% solids is called as *Heavy Tomato Puree.*
- √ *Tomato Paste* is a concentrated tomato juice or pulp free from skin and seeds, containing not less than 25% of tomato solids. *Concentrated tomato paste* contains 33% tomato solids.
- √ *Tomato cocktail is a* product prepared by adding spices including chilli, cloves, cardamom, coriander alongwith vinegar and common salt to the tomato juice. Lemon or lime juice etc may also be added in different proportions to suit the palate.
- √ *Ketchup* is a product made by concentrating tomato juice or pulp without seeds and skin with added, spices, salt, vinegar, onion, garlic etc so that it contains not less than 12 % tomato solids and generally 28% or more total solids (not less than 25% TSS as per FPO specifications).
- √ Ketchup and sauce are similar products. Ketchup is only and only prepared from tomato while sauce may also consist of other pulps i.e. chilli, pumpkin etc.
- √ 1/3rd of sugar is added initially in order to fix the red colour in tomato. Rest of the sugar is added a little before the ketchup is ready.
- √ Tomato soup is a product having min TSS of 7° Brix, which is served hot and is relished equally before principal meals of the day as well as during any other time.

8

Pickles, Fruit Chutney and Sauces

Pickles known to be good appetizers, aid digestion and pallate are edible products preserved and flavoured in a solution of common salt and vinegar. Spices and oil may also be added. There are a number of pickles very popular in India i.e. mango pickle, cauliflower pickle, lime and lemon pickles, ginger chilli pickle etc. But, some pickles like *Sauerkraut* are not very common in India but are very popular in European countries.

Principles of Preservation

- Conversion of fermentable carbohydrates into organic acids (chiefly lactic acid).
- Preservative action of added, vinegar and other ingredients.
- Antimicrobial activity of organic acids, oils, spices etc.

Pickling Process

2 stages

1. Curing or fermentation with dry salting or fermentation in brine or salting without fermentation
2. Finishing and packaging

The lactic acid bacteria are salt tolerant and they can flourish in a brine of about 8-10% salt concentration. The various fruits and vegetables which undergo lactic acid fermentation are cabbage, cauliflower, broccoli, carrot, turnips, radish, cucumber, olive, garlic, apples, pear, immature mangoes, immature plums, lemons, banana etc.

Lactic Acid Fermentation

$$\text{Carbohydrates} \xrightarrow{\text{Homofermentative}} \text{Lactic Acid}$$

$$\text{Carbohydrates} \xrightarrow{\text{Heterofermentative}} \text{Lactic Acid, Acetic acid, } CO_2\text{, Ethanol}$$

Gherkins : Small sized cucumbers brined for pickling

Sauerkraut : Fermented cabbage: It is a sound, clean product of characteristic flavour obtained by full fermentation, chiefly lactic of properly prepared and shredded cabbage in the presence of not less than 2% and not more than 3% salt. It contains upon completion of fermentation, not less than one and a half per cent acid expressed as lactic acid (Frazier and Westhoff, 1995).

TYPES OF PICKLES

1. Salt Pickles

Salt in concentrations above 12 % acts as preservative. It also improves the taste and flavour of pickles.

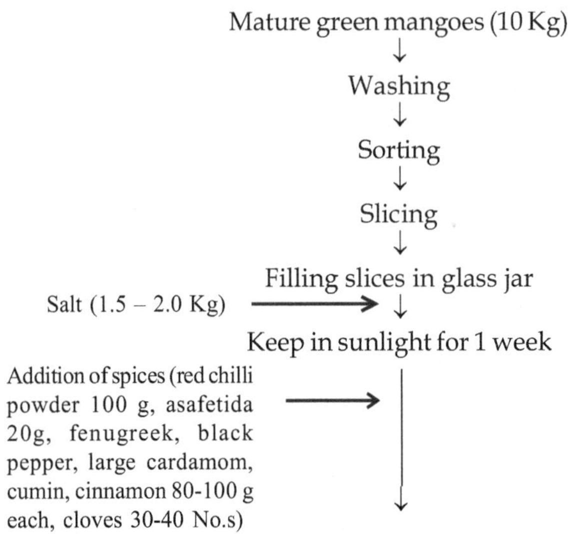

Preparation of Mango Pickle (preservation in salt)

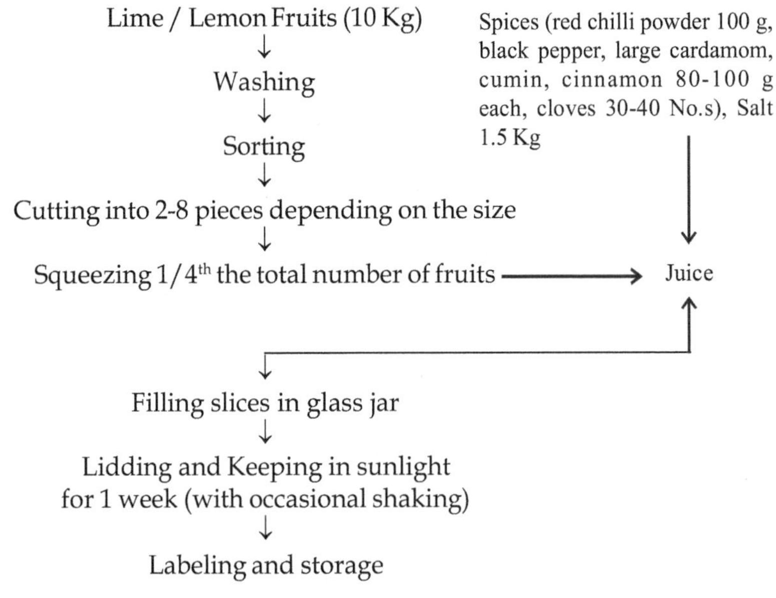

Preparation of Lime / Lemon Pickle (preservation in salt)

2. Oil Pickle

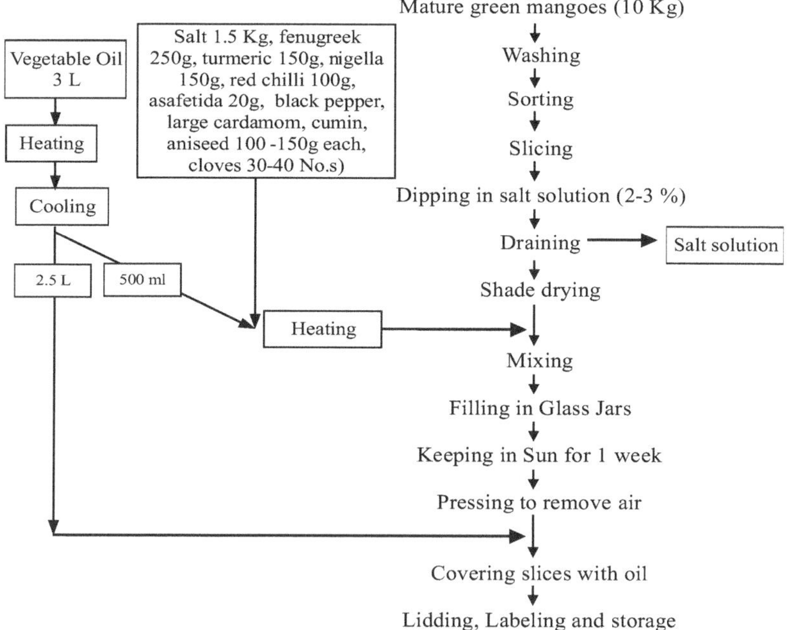

Preparation of Mango Pickle (preservation in oil)

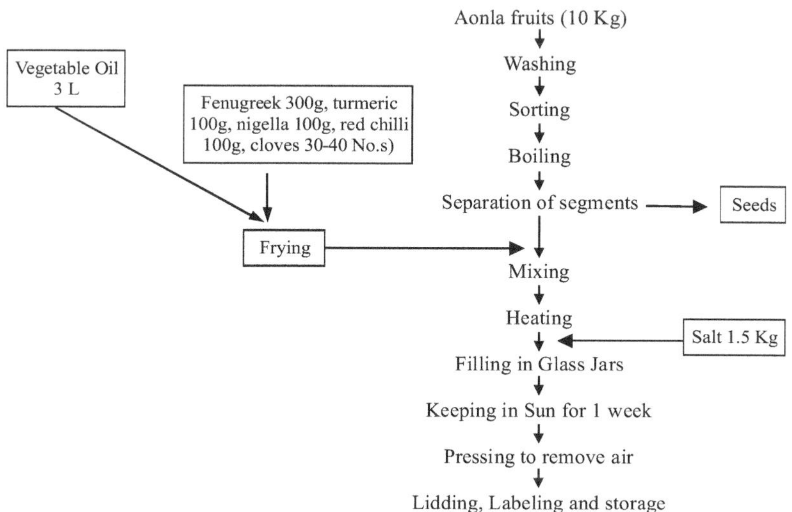

Preparation of Aonla Pickle (preservation in oil)

3. Pickles in vinegar

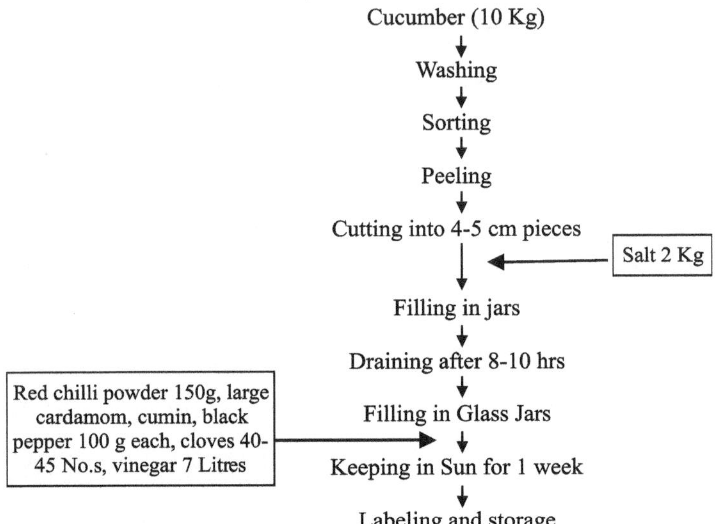

Preparation of Cucumber Pickle (preservation in vinegar)

4. Pickles in mixture of salt, oil and vinegar

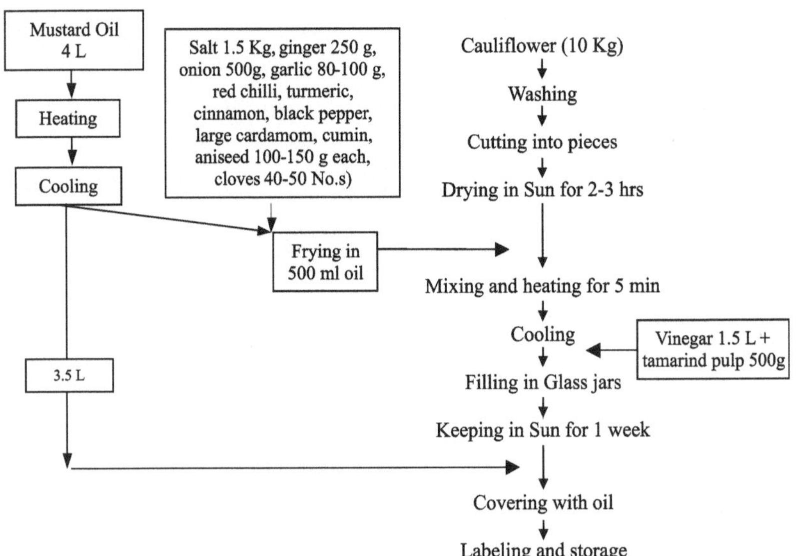

Preparation of Cauliflower Pickle

FPO Specifications

1. Pickles in vinegar : not less than 2% acidity in fluid portion as citric acid.
2. Pickles in citrus juice or brine : not less than 12% salt
   not less than 1.2% acid if the pickle is made in citrus juice
3. Oil Pickles : Any edible vegetable oil like mustard, olive oil etc should be used

Problems in Pickles

Problem	Cause	Solution
1. Spoilage in storage	1. Lower salt, or vinegar concentration 2. Inadequate heating 3. Uncovered slices on the top portion of jar	Use appropriate salt and vinegar concentration. Ensure proper drying and heating of the material. If necessary preservatives may be added.
2. Bitter taste	1. Strong vinegar 2. Excess of spices and prolonged cooking of spices	Use vinegar of proper strength and avoid excess cooking of spices
3. Shrivelling	1. Putting vegetables i.e. cucumber in very strong brines.	Use lower strength brines and increase the concentration gradually
4. Scum formation	1. growth of wild yeast on the surface of brine during curing	Remove scum as soon as possible. Add 1% acetic acid to the brine for preventing growth of wild yeast
5. Slipperiness and Softness	1. Use of low strength brines 2. inadequate covering with brine	Use brines of proper strength and cover the material completely with brine.
6. Cloudiness in vinegar	1. Inadequate penetration of vinegar into vegetable tissues 2. Fermentation starts from inside the tissues	Use brines of proper strengths Perform proper curing Use vinegar of good quality
7. Blackening	1. Use of iron equipment during handling and processing	Use stainless steel equipments
8. Colour change, darkening, fading	1. Insufficient curing 2. Improper packaging 3. Exposure to air	Perform proper curing and proper packaging. Exclude all the air as possible.

CHUTNEY

It is a product prepared by cooking the fruit pulp with added salt, sugar, spices, acetic acid and/or dry fruits to a suitable consistency so that it contains TSS not less than 50% and fruit part not less than 40% a per FPO. Sugar, salt, spices, acetic acid all act as partial preservatives. Preservatives may also be added in chutney for increasing its storability. Generally no artificial colour and flavour are required to be added.

A general Recipe for the Preparation of Fruit Chutney

Fruit pulp (TSS 10-15° Brix and acidity about 1%)	5 Kg
Sugar	2.4-2.6 Kg
Salt	140-150g
Red chilli powder (according to taste)	15-25 g
Black pepper	35-40 g
Cardamom	70-80g
Cumin	40-50 g
Ginger chopped	350g
Citric acid	15-20g
Acetic acid	15-20 ml
Sodium Benzoate	1 g

SAUCE

Sauce is a product similar to ketchup, prepared from pulps of tomato or other fruits/vegetables, having TSS not less than 15% and cooked to a suitable consistency with added sugar, salt, spices and vinegar (acetic acid). Sugar, salt, spices, acetic acid all act as partial preservatives. According to the FPO fruit sauce should have a minimum of 15% TSS and 1.2% acidity. Preservatives and colours may also be added in sauces for increasing its storability. Sauces may or may not be prepared from tomato, but ketchups are essentially prepared from tomato. Some examples of sauces are tamarind sauce, pumpkin sauce, chilli sauce, soya sauce etc. The procedure for preparation of sauce is similar to as that of ketchup as discussed in previous chapter.

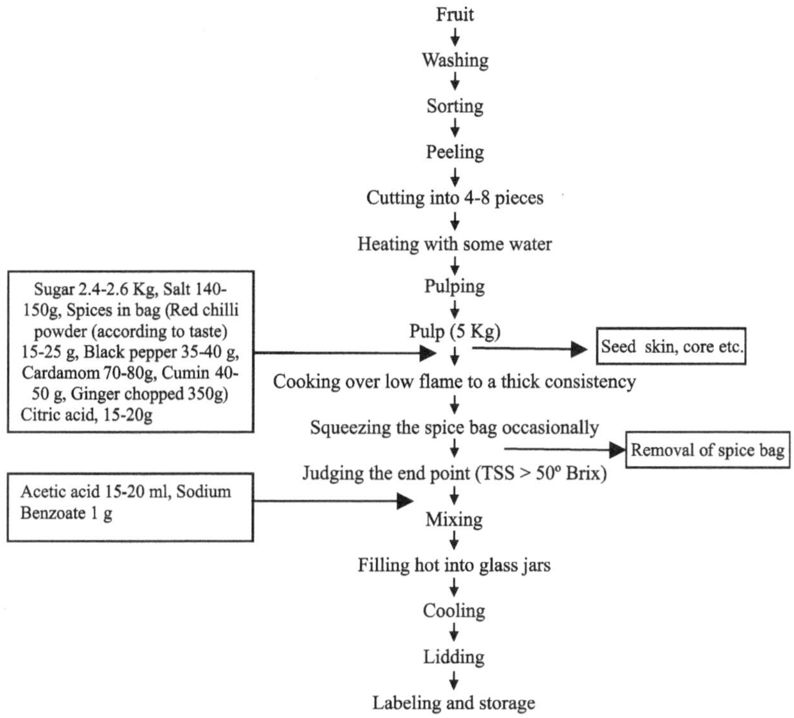

Preparation of Fruit Chutney

Black neck is an important problem encountered during preparation of sauces and ketchups. In this a black ring is formed in the neck portion of the sauce / ketchup bottle in storage. This is caused due to the incorporation of iron into the product.

Iron (Fe) *(crown corks/ Equipment)* + Tannins *(Spices)* ⟶ Ferrous tennate ⟶ Ferric tennate *(Black colour)*

Remedy

- Do not use iron vessels for cooking spices
- Add 100 ppm SO_2

- Leave less head space (air) while packing
- Store the bottles inverted / horizontal so that air diffuses throughout the product and the availability in the neck portion is reduced.
- Remove the flower head of cloves when used.

Points to Remember

√ Pickles known to be good appetizers aid digestion and add to the pallate are edible products preserved and flavoured in a solution of common salt and vinegar. Spices and oil may also be added.

√ Conversion of fermentable carbohydrates into organic acids; addition of sufficient amount of sugar, vinegar and other ingredients to the fully cured and packed product; antimicrobial activity of organic acids, oils, spices etc. are the important principles of preservation used in pickles

√ The lactic acid bacteria are salt tolerant and they can flourish in a brine of about 8-10% salt concentration.

√ Homofermentative bacteria produces only lactic acid while heterofermentative bacteria produces lactic acid, acetic acid, CO_2, ethanol etc.

√ Gherkins are small sized cucumbers brined for pickling

√ Sauerkraut is fermented cabbage pickle. It is a sound, clean product of characteristic flavour obtained by full fermentation, chiefly lactic of properly prepared and shredded cabbage in the presence of not less than 2% and not more than 3% salt.

√ Pickles can be preserved in salt, vinegar, oil or mixture of the three

√ Shrivelling in pickles is caused due to use of high strengths brines during curing, while bcackening is caused due to use of iron equipments during handling and preparation

√ Chutney is a product prepared by cooking the fruit pulp with added salt, sugar, spices, acetic acid and / or dry fruits to a suitable consistency. Sugar, salt, spices, acetic acid all act as partial preservatives. According to the FPO fruit chutney should have a minimum of 50% TSS and 40% fruit part.

√ Sauce is a product consisting of fruit pulp cooked to a suitable consistency with added sugar, salt, spices, vinegar (acetic acid). Sugar, salt, spices, acetic acid all act as partial preservatives. According to the FPO fruit sauce should have a minimum of 15% TSS and 1.2% acidity.

√ *Black neck* is a problem of sauces and ketchups in which a black ring is formed in the neck portion of the sauce / ketchup bottle in storage. This is caused due to the incorporation of iron into the product and its reaction with tannins of the spices.

9

Jam, Jelly, Marmalade and Preserve

Jam : It is a product prepared by boiling to a suitable consistency, fruit pulp with sufficient quantity of sugar, acid and pectin so as to set and become firm enough to hold the tissues in position. About 45 parts of pulp are used with every 55 parts of sugar. The finished product should not contain less than 68 per cent total soluble solids.

Jelly : Jelly is prepared by boiling the fruit pectin extract with sugar in the presence of appropriate amounts of acid so as to set into a clear gel. A perfect jelly is transparent, devoid of haziness or any suspended material well set, but not too stiff and should have original flavour of the fruit. When cut it should retain its shape and show a smooth cut surface.

Marmalade : It is a fruit jelly like product in which slices of fruit or peel are suspended in the set gel. Mostly they are prepared from citrus fruits like oranges, lemons etc.

Preserve : It is a product prepared by keeping and preserving the fruit as whole or in large pieces in sugar syrup until the soluble solid concentration of about 68% or more is achieved.

Principle of Preservation in Jam, Jelly, Marmalade and Preserve

- Sugar under high concentration (>66%), acts as a preservative.
- The moisture present becomes bound and is not free for the growth of microorganisms.
- The water activity is reduced creating unfavourable conditions for the growth of microorganisms.

Difference between Jam and Jelly

Jam	Jelly
1. Prepared from fruit pulp	1. Prepared from clear fruit pectin extract
2. Not transparent	2. It is transparent
3. Min. TSS 68 % as per FPO specifications	3. Min. TSS 65 % as per FPO specifications
4. Flavours and colours are generally added	4. Should have the original flavour of the fruit.
5. Opt. pH 3.35 - 3.70	5. Opt. pH = 3.30

Preparation of Jam

1. *Preparation of pulp* : The fruits after thorough sorting and washing are prepared for extraction of pulp. Fruits, like plums, apricots, peaches, apple etc. are heated with about 10 % water until they become soft and passed through the pulper to separate the skin and stones etc. In case the jam is being prepared from the pulps which have been preserved in 1000-1500 ppm SO_2, the SO_2 needs to be

removed from the pulp before product preparation. So the preserved pulps are boiled with addition of small quantities of water till the SO_2 is not perceived.

2. *Addition of sugar* : Cane sugar in the proportion of 45 :55 (fruit part : sugar) is added during boiling. Finished jam should contain 30-50% of invert sugar or glucose to avoid crystallization of cane sugar during storage. But if the reducing / invert sugars exceed 50% the jam will develop into a heavy honey like syrupy mass.

3. *Addition of acids* : The acidity of the finished product varies between 0.5 to 0.7 %. So the acid may be added accordingly if needed.

4. *Processing* : The cooking or processing may be done in SS kettles or steam jacketed kettles. Vacuum evaporators, plate evaporators may also be used for processing.

5. *Addition of pectin* : One part of pectin is mixed with about 10 parts of sugar to disperse it uniformly and then is added slowly in the boiling mixture normally after the jam attains 60°Brix, Jam should contain 0.5-1.0% pectin.

6. *Addition of colours and essences* : These are added towards the end of boiling / product preparation. Colour should be dissolved in minimum quantity of water and poured drop wise over the almost finished jam with constant stirring.

7. *End point determination* : Although the end point can be judged easily with the increase in experience, but for the beginners the following methods can be used.

   - *Drop test* : This method can be used where no facility for any type of measurement is available. A small quantity of product is taken in a spoon during cooking and is allowed to air cool for some time. One drop of the air cooled jam is put into a glass of water. If the drop settles down at the bottom of the glass without disintegrating, the product is ready. If it disintegrates and splits apart, it may be cooked for some more time.

- *Boiling point* : Jam is assumed to be ready when the temperature of the boiling mixture reaches 106° C at sea level. The temperature may be a little lesser at high altitudes.
- *TSS* : At end point the TSS of the jam is above 68°Brix. Care should be taken to cool the drop of material which is put on the prism of refractometer. High temperatures may lead to condensation of vapours and thus giving incorrect readings.
- *Weight test* : At the end point, the weight of the jam is about 1.5 times the weight of the added sugar. For using this method the weights of the boiling pan and sugar are to be noted down before starting cooking. The boiling mixture is weighed alongwith the pan for determination of end point.

8. *Filling and packaging* : Prepared jam is poured hot (>85° C) into the jars. Sometimes molten paraffin wax is put on the jam surface. 40 ppm SO_2 may be added for preservation.

Calculations

Suppose that the apple pulp has TSS of about 15 °Brix and acidity of 1.2 % and you wish to prepare apple jam. The following calculations shall be applicable.

Apple pulp	2 Kg
Sugar to be added	2.4 Kg (Pulp and Sugar are added in ratio of 45 :55)
Expected quantity of jam to be prepared	3.6 Kg (Weight of final Jam is about 1.5 times the weight of sugar added)
Suppose the acidity of the finished products is 1 %.	
Quantity of acid required in the finished product	36g

contd...

Quantity of acid contributed by the pulp	1.2 x 10 x 2 = 24 g
Quantity of citric acid to be added	36 - 24 = 12 g

Note : Some weight loss in the finished product may be due to evaporation and some wastage in the container or cooking pan. This is significant when product is prepared in small quantities. The weight loss is very sharp when product is concentrated beyond 70° Brix.

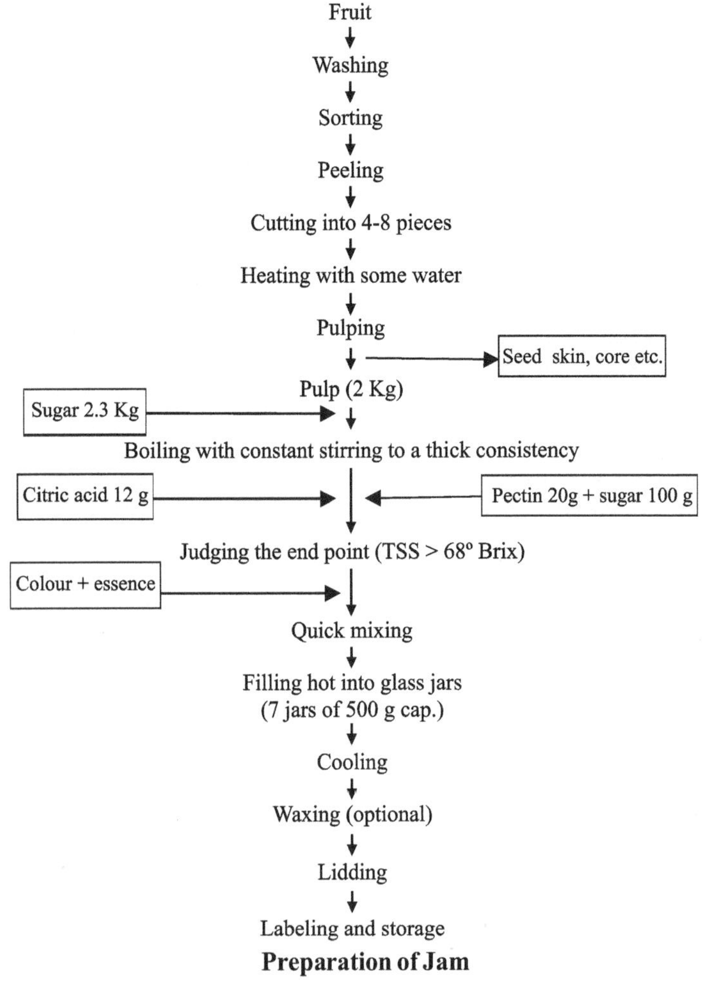

Preparation of Jam

Problems in Jam

Problem	Cause	Remedy
Crystallization	Improper inversion of sugar	Add corn syrup or glucose alongwith cane sugar
Stickiness / gummyness	Higher TSS and higher inversion of sugar and lesser pectin	Add proper proportion of pectin
Premature setting	High pectin and low TSS	Add adequate sugar
Microbial spoilage	High RH, and moisture on surface	Add SO_2 @ 40 ppm

Preparation of Jelly

Jelly is a well set transparent semi-solid product prepared from fruit pectin extract on cooking with added sugar and acid. 4 essential constituents of jelly (finished product) are :

1. Pectin 0.5 - 1.0 % (normally obtained from fruit)
2. Sugar 60-65 %
3. Fruit acid 1.0 % (min. 0.5 %) (opt. 0.75%)
4. Water 33-38%

The middle lamella of plant cells consist of protopectin which is insoluble in water. The protopectin when boiled in the presence of acids is hydrolysed to water soluble pectin. Pectin particles are negatively charged and a pectin solution is most stable in neutral pH range. Increase in acidity or alkalinity decreases its stability. In jelly formation, sugar acts as a precipitating agent and the presence of acid helps it. More the quantity of acid, the less sugar is required. Some salts also help in precipitating pectin.

Selection and Choice of Fruits

Fruits should be rich in pectin and acid eg guava, apples (sour and crab), lemons, oranges etc. If the pectin or acid is deficient, it may be supplemented with addition of pectin or citric acid or by addition of pectin or acid rich fruits. Both ripe and under ripe fruits should be used as the former yields good flavour and the latter good pectin. Over ripe fruits may be avoided as they may disintegrate during boiling and affect the clearity of extract adversely.

Extraction of Pectin

The fruits are cut into pieces of about 1/8th to 1/4th of an inch in thickness and used for pectin extraction. About ½ to an equal volume of water is added to the fruits in case of apples, while citrus fruits need two to three volumes of water for each volume of sliced fruit due to prolonged boiling required. However, fruits like grapes and berries do not require any addition of water. Time of boiling is about 20-25 min for apple, 5-10 min for grapes and other berries and about 45-60 min for oranges, 30-35 min for guava. Boiling should be done in stainless steel containers.

Straining and Clarification of Extract

The boiled fruit mass is then strained through cloth or cheese cloth folded several times. The bag containing fruits is not squeezed to prevent the solid particles to pass through the cloth.

Determination of Pectin Content of the Extract

a. Alcohol Test

1 spoon full of extract (cooled) + 3 spoon full of methylated spirit

↓

Mix and keep undisturbed for 5 min.

↓

Observe and interpret as follows

Type of clot formed	Pectin content	Recommended extract : sugar ratio
Single transparent lump	Rich	1.00 : 1.00
Less firm and fragmented	Moderate	1.00 : 0.75
Numerous small granular clots	Poor	1.00 : 0.50

b. Jelmeter Test

Pectin extract is allowed to flow through the jelmeter (on which graduations are marked as ½, ¾, 1 and 1¼) exactly for 1 min. The level upto which the material has moved indicate the ratio of extract and sugar to be added as 1 : ½, 1 : ¾, 1 : 1 and 1 : 1¼.

c. Making Test Jellies

The quantity of pectin extract is kept constant and quantity of sugar is varied. Test jellies are prepared with all the combinations. The proportion that gives the best jelly is selected for preparation at larger scales.

Addition of Sugar and Acid

Exact quantity of sugar to be added is calculated based on the above three tests. Sugar that is to be added should be clean and not containing any dirt etc otherwise it may reduce the transparency of the final product. The optimum pH for jelly containing 60, 65 and 70 % sugar is approximately 3.0, 3.2 and 3.4 respectively. Acid may be added to the concentrated mass just prior to filling into the containers.

Boiling

Prolonged boiling results in the formation of syrupy jelly and darkens its colour. To limit the time of boiling pectin extract should be brought to boil first and then sugar added to it. During boiling, the scum which rises to the top is removed from time to time. Foaming can be avoided by adding 1 table spoon of edible oil per 50 Kg boiling jelly. For proper inversion, the boiling should be completed within about 20 min.

End Point Determination

The end point in case of jelly and marmalade can be judged by the following methods

1. *Boiling point* : Sugar solution containing 65% sugar boils at 105° C at sea level. The same shall hold true for jelly, but temperature corrections are necessary depending upon the altitude of the place. Generally the end point of jelly is 3-4° C higher than the boiling point of water at that place.

2. *Sheet or ladle test* : Take some boiling jelly in a ladle and cool it slightly and allow it to drop into the pan. If it drops like syrup, the mixture requires more cooking and if it falls like a sheet or flake, the end point has been reached.

3. *Cold plate test* : Take a drop of boiling mixture and place it on a cold inverted steel plate. Cool it rapidly and push it slightly with a finger. If the jelly is ready, the cooled drop will crinkle and come all at once without breaking.

4. *Weight test* : At the end point, the weight of the jelly should be 1.5 times the weight of the added sugar. For using this method the weights of the boiling pan and sugar are to be noted down before starting cooking. The boiling mixture is weighed alongwith the pan for determination of end point.

Packing : Jelly is poured at about 85° C into glass jars and allowed to cool rapidly to room temperature. Any foam on the top of jar is removed by spoon.

Problems in Jelly

Problem	Cause	Remedy
i. Failure of jelly to set	a. Lack of acid or pectin b. Addition of too much of sugar c. Inaccurate measurement d. Insufficient cooking e. Overcooking or re-cooking	a. Addition of sufficient quantity of freshly extracted pectin extract b. Addition of appropriate quantities of acid c. Proper cooking upto end point
ii. Cloudy jelly	a. Unclarified juice or pectin extract b. Under-ripe fruit c. Non-removal of scum d. Premature gelation e. Overcooking f. Slow pouring into jars	a. Use properly clarified extract b. Remove the scum quickly c. Do not pour jelly from too much height
iii. Colour changes	a. Storing in too warm place b. Imperfect jar seal c. Trapped air bubbles causing oxidative changes	a. Seal the jars properly b. Store at a cool place
iv. Crystal formation	a. Excess of sugar b. Overcooking c. Tartrate crystals in grape jelly if the juice is kept cool for several hrs before use	a. Add proper quantity of sugar b. Do not overcook
v. Gummy jelly	1. More inversion of sugars due to overcooking	a. Ensure proper inversion of sugar
vi. Weeping jelly/ Syneresis	1. Overcooking 2. Excess of acid 3. Too low sugar 4. Insufficient pectin	a. Add proper pectin and sugar b. Do not add excess acid c. Store at a cool dry place
vii. Stiff jelly	1. Overcooking 2. Excess of pectin	a. Add proper quantity of pectin and avoid overcooking
viii. Mould growth	1. Imperfect seal 2. Air borne contamination 3. Under processing	a. Cook the jelly upto proper TSS b. Seal properly in glass jars

* *Syneresis* : Phenomenon of spontaneous exudation of fluid from the gel.

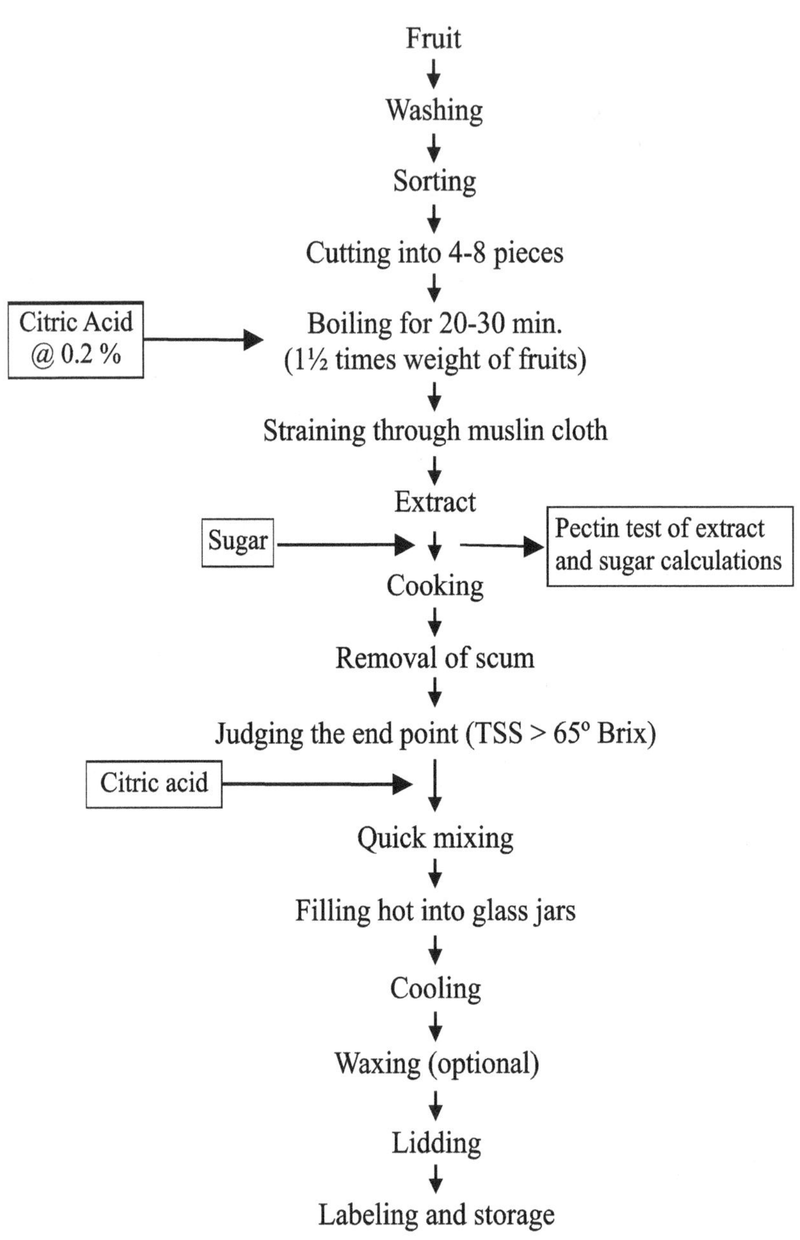

Preparation of Jelly

Preparation of Marmalade

It is a jelly like product made commonly from citrus fruits with shreds of peel suspended in the set gel. Pectin and acid levels are generally kept on a higher side than that in jelly. Bitterness imparted by peel shreds is regarded as a desirable characteristic of marmaldes.

Preparation of Extract

The outer coloured portion of skin (flavedo) is peeled off. The albedo portion, which is rich in pectin is boiled alongwith the cut slices for about 45-60 minutes in 2-3 times of fruit weight in water. The extract is separated and filtered using filter aids. The extract should have at least 1 % acidity as citric acid.

Preparation of Peel Shreds

The large pieces of the flavedo are cut into slices of about 2-3 cm length and 1 cm thickness and then are softened by boiling in water for 10-15 minutes with added pinch of tartrazene colour to give shine to shreds or by autoclaving at 10-15 lb/m^2 pressure for a few minutes.

The shreds may be kept in heavy syrup for sometime before addition to jelly to increase their bulk density and to avoid floating on the surface.

Boiling

The fruit extract is first boiled before addition of sugar. When temperature of mixture reaches 103^0C at sea level the prepared shreds are added @ 5-7 per cent of the original extract. Boiling is done till it reaches end point which is determined the same way as in case of jelly.

Cooling

Prepared marmalade is allowed to cool for some time with occasional stirring for uniform distribution of peel shreds till temperature reaches 85^0C and a thin film begins to form on the surface that prevents shreds from coming to the surface.

Flavouring

A few drops of orange oil are mixed in marmalade before filling into containers.

Packing

In similarity with jams and jellies, marmalade is also filled into glass jars at about 85°C.

Storage

To avoid darkening and mould growth during storage 40 ppm of SO_2 may be added.

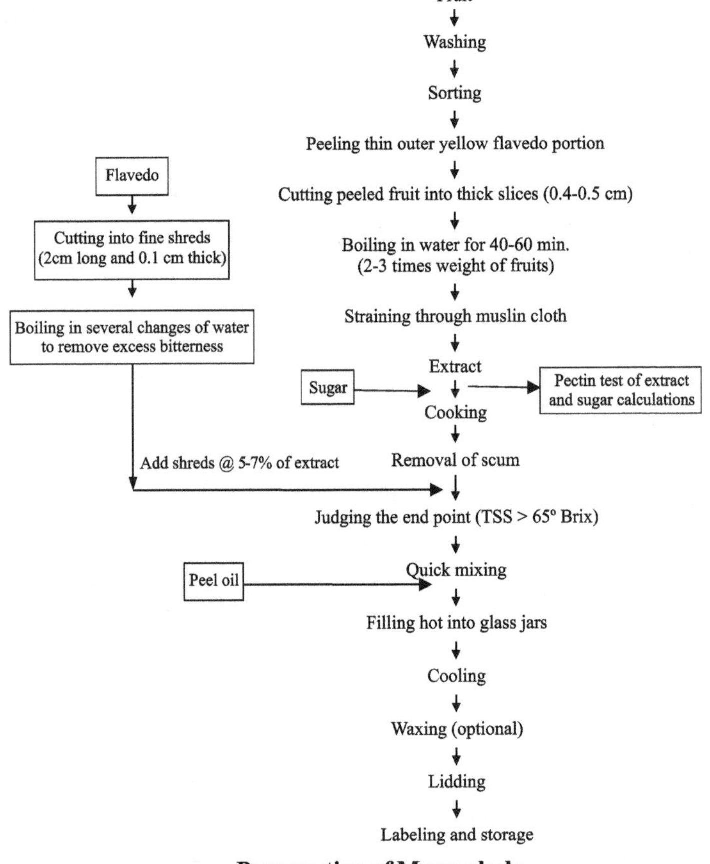

Preparation of Marmalade

Preserves, Candied and Crystallized Fruit Products

Preserve : Preserve is a product prepared by keeping and preserving the fruit or vegetable as whole or in large pieces in sugar syrup until the soluble solid concentration of about 68% or more is achieved. The fruit should retain its form, crispness and tenderness without shriveling of individual pieces. eg aonla preserve, apple preserve, carrot preserve, mango preserve.

Methods of Preparation

i. *Rapid method* : Fruits are cooked in low concentration sugar syrups over gentle flame till the syrup becomes thick and TSS reaches beyond 68° Brix. Soft fruits like strawberries and raspberries can be processed by this method. This is not very common as the process may make the fruits tough.

ii. *Slow method* : Fruits are peeled and blanched as the case may be and kept in alternate layers of equal quantity of sugar as the fruit weight for 24 hours. The sugar penetrates into the fruits and water exudes out and dissolves the rest of the sugar. Next day material is drained over a sieve to separate the syrup and fruit pieces. Syrup is heated to concentrate it upto 60° Brix and fruit pieces are put back into the concentrated syrup. Next day the same process is repeated and syrup is concentrated to 65° Brix. Next day again the same process is repeated and sugar syrup is concentrated to 70° Brix and fruit pieces are finally packed glass jars and preserved in concentrated syrup of 70° Brix. Product is kept for few days before consumption.

Candied fruits/ vegetables : It is a product prepared by impregnation of sugar into the fruit /vegetable pieces followed by surface drying to give a transparent coating of sugar on the surface. The method of preparation of candied fruits is also similar to that of preserves except that the excess syrup is drained off and the pieces are surface dried. The drained syrup can be used for the preparation of next batch of product or in pickles, sauces, vinegar making etc.

Crystallized fruits / vegetables : When candied products are covered with sugar crystals by rolling and allowing crystals to settle on the surface it is known as *crystallized fruits / vegetables*.

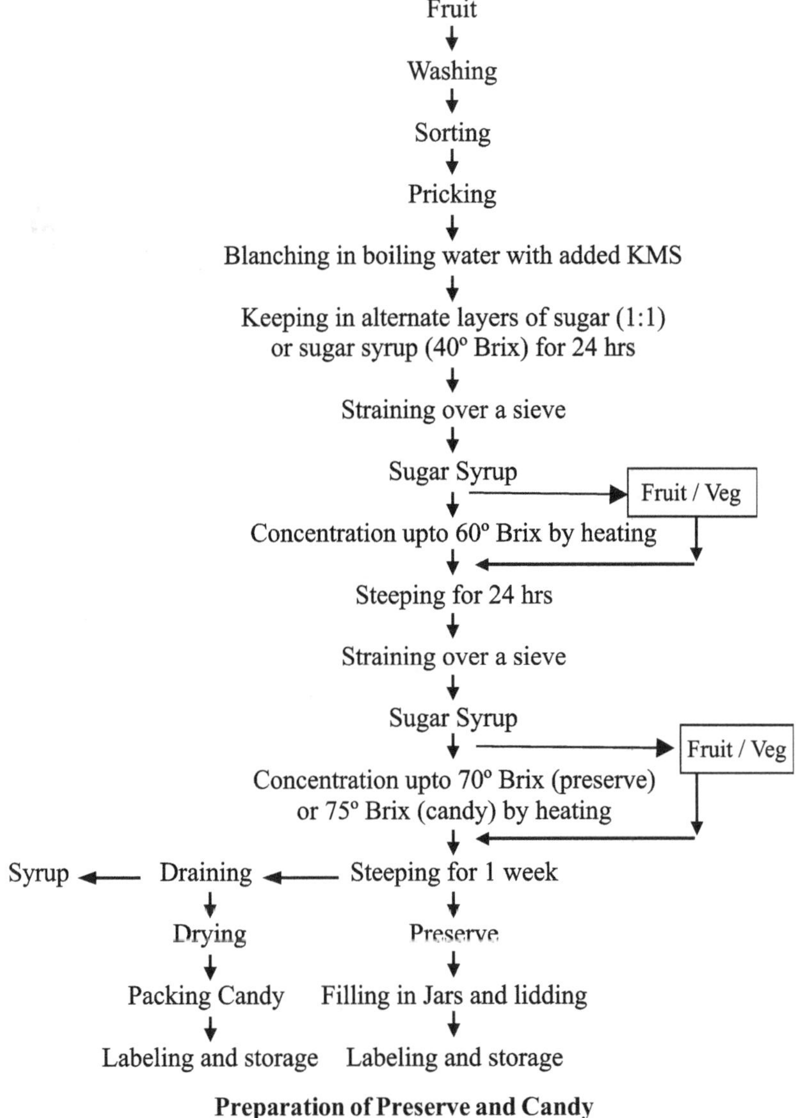

Preparation of Preserve and Candy

Problems in Preserves and Candy

Problem	Cause	Remedy
1.Fermentation	1. Low sugar concentration 2. Inadequate cooking	1. Add sugar 2. Concentrate to 70-75° Brix 3. Store at a cool dry place
2. Floating fruits in syrup	1. Inadequate impregnation of sugar into the fruit 2. Filling the preserve without cooling	1. Ensure raising of specific gravity of fruits by proper sugar impregnation 2. Cool the preserve before filling into jars
3. Toughning and hardening of fruit	1. Inadequate blanching 2. Cooking in large shallow pans in little quantity of syrup	1. Ensure proper blanching 2. Cook in sufficient syrup
4. Shrinkage of fruits	1. Cooking directly in high concentration syrups	1. increase the sugar strength gradually
5. Stickiness	1. Insufficient drying of candy 2. Storage in RH 3. Poor packaging	1. Dry the candy properly 2. Pack properly and store at a cool dry place

Points to Remember

√ Jam is a product prepared by boiling to a suitable consistency, fruit pulp with sufficient quantity of sugar, acid and pectin so as to set and become firm enough to hold the tissues in position. About 45 part of pulp are used with every 55 parts of sugar.

√ Jelly is prepared by boiling the fruit pectin extract with sugar in the presence of appropriate amounts of acid so as to set into a clear gel. A perfect jelly is transparent, devoid of haziness or any suspended material well set, but not too stiff and should have original flavour of the fruit. When cut it should retain its shape and show a smooth cut surface.

√ Marmalade is a fruit jelly like product in which slices of fruit or peel are suspended. Mostly they are prepared from citrus fruits like oranges, lemons etc.

- √ Preserve is a product prepared by keeping and preserving the fruit as whole or in large pieces in sugar syrup until the soluble solid concentration of about 68% or more is achieved.
- √ Reduction of water activity and the fact that sugar acts as preservative above 66% concentration are the main principles of preservation of jam, jelly, marmalade, preserve, candy and crystallized fruits.
- √ Jam is prepared from fruit pulp while jelly from clear fruit pectin extract
- √ The TSS in case of jam is 68 and jelly 65 preserve 68 and candy 75
- √ Finished jam should contain 30-50% of invert sugar or glucose to avoid crystallization of cane sugar during storage. But if the reducing / invert sugars exceed 50% the jam will develop into a heavy honey like syrupy mass.
- √ The acidity of the finished jam varies between 0.5 to 0.7 %.
- √ Boiling point of jam and jelly is 105-106 °C at end point
- √ Weight of finished jam and jelly is 1½ times the weight of added sugar in general.
- √ Protopectin is water insoluble and when boiled in the presence of acids it is hydrolysed to water soluble pectin.
- √ Guava, apples (sour and crab), lemons, oranges etc. are rich in pectin and are suitable for preparation of jelly.
- √ Both ripe and under ripe fruits should be used as the former yields good flavour and the latter good pectin.
- √ During preparation of extract the bag containing fruits is not squeezed to prevent the solid particles to pass through the cloth.
- √ Alcohol test and jelmeter test are used for measuring strength of pectin
- √ Weeping jelly/ Syneresis is the phenomenon of spontaneous exudation of fluid from the gel

- √ Jelly may fail to set due to lack of acid or pectin, addition of too much of sugar, inaccurate measurement, insufficient cooking and overcooking
- √ Bitterness is regarded as a desirable characteristic of marmalades
- √ Marmalades are prepared from citrus fruits
- √ Candied fruits/vegetables are the product prepared by impregnation of sugar into the fruit/vegetable prices followed by surface drying to give a transparent coating of sugar on the surface.
- √ When candied products are covered with sugar crystals by rolling and allowing crystals to settle on the surface it is known as *crystallized fruits/vegetables*.

10

Canning of Fruits and Vegetables

Canning : Process of preservation of various food stuffs including fruits and vegetables whole or in pieces, in sugar syrup or brine by heat processing them in hermetically sealed containers.

History

- 1804 : Nicholas Appert, a Persian confectioner (known as Father of Canning) invented the process of preserving foods in hermatically sealed glass containers by application of heat. Canning is also called as *Appertizing or Appertization.*
- 1807 : Method of canning foods was described by Thomas Saddington in England.
- 1810 : Peter Durand obtained British patent on canning of foods in tin containers.
- 1817 : William Underwood introduced canning on commercial scale in USA.

- 1860s : Louis Pasteur proved that microorganisms are the real cause of spoilage and by destroying them food can be preserved in suitable containers. He introduced the term *Pasteurization* which refers to *"preservation of food by heat treatment at sufficiently high temperature below 100° C to kill majority but not all of the microorganisms and by preventing their access to the food inside containers by sealing it hermatically"*.

Requirements of Fruits and Vegetables for Canning

Fruits

1. Should be ripe but firm and evenly matured
2. Should be free from blemishes, insect damage and malfomation
3. Should not be over ripe as these may be infected with micro-organisms and loose their texture and integrity during heat processing
4. Under ripe fruits are avoided as they shrivel and toughen on canning.

Vegetables

1. Should be tender
2. Should be free from soil, dirt etc.
3. Tomatoes should be firm, fully ripe and deep red in colour.

Steps in Canning of Fruits and Vegetables

1. *Sorting and Grading* : For obtaining a pack of uniform quality as regards the size, colour etc., proper sorting and grading by hand or grading machines is essential.
2. *Washing* : Fruits are washed with cold water or hot water sprays or by soaking or agitation. Vegetables may be soaked in dilute solution of $KMnO_4$ to disinfect them.

3. *Peeling, Coring etc.*
   a. *By hands* : All fruits and vegetables
   b. *By machine* : Abrasive peelers (potato, carrot, turnips etc.)
   c. *By Heat* : Some fruits and vegetables like peaches and potatoes are scalded in steam or boiling water or at high temperature of 40° C for about 1 min. followed by removal of skin by hands or pressure sprays.
   d. *Lye Peeling* : It is a process by which commodities like peaches, apricots, quinces, carrots, sweet potatoes etc are peeled by dipping them in boiling caustic soda (NaOH) or lye solution of 1-2 % for 30 sec to 2 min. depending upon the type and maturity of the commodity. Peel is then removed by hands or water sprays. The fruits are washed in water or dilute solution of HCl to remove excess of alkali. So *lye peeling* is the process of removal of peel in boiling caustic soda solution. Use of aluminium equipment should be avoided during lye peeling as aluminium reacts with NaOH.
4. *Blanching* : Treatment of fruits or vegetables (whole or in pieces) with boiling water or steam for short periods of time followed by immediate cooling prior to canning done primarily to inactivate enzymes is called blanching.

Functions

(i) Inactivates the enzymes thus preventing the possibility of discolouration.

(ii) Helps cleaning the fruits and vegetables (remove dirt, soil etc.).

(iii) Reduces microbial load.

Why cooling is necessary after blanching?

(i) To avoid over-cooking

(ii) To retain texture and integrity of fruits and vegetables

(iii) To give temperature shock to the microorganisms

5. *Can Filling :* Cans are first washed with hot water or steam jets to remove any dirt or dust. In India, fruits are filled in the cans by hands (using rubber gloves) to ensure proper grade and to prevent any bruising.
6. *Syruping and Brining :* Generally cans are filled with hot sugar syrup for fruits and hot brine for vegetables. The strength of syrup is measured with refractometer while that of brine with salometer or salinometer or Baume' hydrometer.

100° Salometer reading = 26.5 % salt concentration (maximum solubility of salt in water is 26.5%).

Objectives of adding syrup or brine :

a. To improve taste of the canned fruit or vegetable
b. To fill up the interspace between fruit or vegetable pieces.
c. To facilitate further heat processing.

The temperature of syrup or brine at the time of can filling should be 79-82° C. Head space of 0.32 to 0.47 cm should also be kept at the time of filling syrup or brine.

7. *Lidding or Clinching :* Placing the lid on the top of the filled can (lidding) and partially seaming the lid to the can by a single *first roller operation* action of the double seamer is called clinching. Lid remains sufficiently loose to permit the escape of free as well as dissolved air out of the can.

8. *Exhausting :* Before sealing the cans finally, by second roller operation of double seamer, it is necessary to remove practically all the air from the contents of the can by the process known as exhausting. So, exhausting is an unit operation done after clinching and before sealing to removed dissolved air from the can contents. The can is passed through a tank of hot water at about 82-87° C for 5-25 min so that the temperatures at the centre of the can becomes minimum 79° C.

9. *Sealing* : Properly exhausted cans are sealed by the *second roller operation* of the double seamer. After sealing air can not enter into the can and cause any kind of spoilage.

10. *Heat processing* : Heating followed by cooling of cans is done to kill bacteria and other microorganisms.

    - In general it is said that almost all fruits can be processed satisfactorily at 100° C in boiling water as the the presence of acids retards the growth of bacteria.

    - Vegetables (except tomato), being non-acidic, are required to be processed at 115-121° C.

    The centre of the can should receive these temperatures and the can should be maintained at these temperatures for sufficient long time to kill all the potential microorganisms and their spores.

11. *Cooling* : After processing the cans are cooled immediately in cold water to room temperature in order to

    a. Stop over-cooking

    b. Prevent stack burning

    c. Give temperature shock to the microorganisms if any

12. *Testing of cans*

    a. *Ringing sound* : A clear ringing sound indicates perfect seal while dull and hollow sound indicates imperfect seal.

    b. Concavity of the lid also gives and idea about the degree of vacuum inside the can.

    c. Vacuum gauge

    d. *Flip tester* : By placing the can inside a glass chamber and evacuating it

13. *Labelling, packaging and storage.*

Standard Sizes of Cans

Trade name	Size in inches	Size in mm	Trade size	Capacity cm^3
A 1	2.11/16 x 4	68 x 102	211 x 400	316
A1 Tall	3.1/16 x 4. 11/16	78 x 119	301 x 411	479
A 2	3.7/16 x 4.8/16	87 x 114	307 x 408	579
A 2½	4.1/16 x 4.11/16	103 x 119	401 x 411	848
A 10	6.3/16 x 7	157 x 178	603 x 700	3069
1 lb jam	3.1/16 x 3.9/16	78 x 90	301 x 309	356
2 lb jam	4.1/16 x 4	103 x 102	401 x 400	721
7 lb jam	6.3/16 x 5.13/16	157 x 148	603 x 513	2543

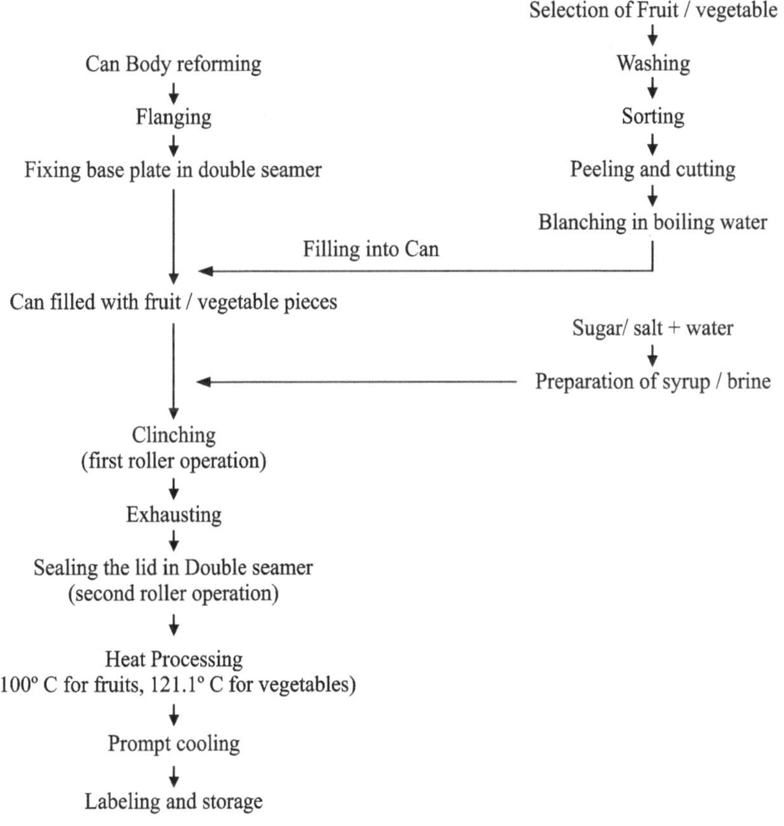

Canning of Fruits and Vegetables

Difference between Clinching and Sealing

Clinching	Sealing
1. Done by first roller operation of double seamer	1. Done by second roller operation of double seamer
2. Done before exhausting	2. Done after exhausting
3. Air can enter the can or come out of the can after clinching	3. Air movement is not possible after sealing

Problems in Canned Foods

Problem	Symptoms	Cause	Remedy
1. Swell	Ends are perfect, can becomes bulged due to positive internal pressure	Gas formation by microbial and chemical action	• Use good quality, uniformly coated tin plates, • Add 0.5 % citric acid to syrup as it checks hydrogen swells, • Lid should be clinched by I roller operation of double seamer before exhausting, • Exhaust properly and seal (II roller operation) at temperatures not less than 74° C • Cool the cans immediately and store at a cool dry place.
2. Hydrogen swell	Ends are perfect, can becomes bulged due to positive internal pressure	Formation of hydrogen gas on reaction of metal and acids in food. Food generally is free from harmful microorganisms	
3. Flipper	Can ends become convex but can be pushed back on applying pressure but they again flip back to their convex position when pressure is released.	Over filling, under exhausting and gas formation due to spoilage.	
4. Springer	Mild swell or bulging at both ends of can. It can be pressed but again springs back and becomes convex after some time.	Initial stage of hydrogen swell or due to over filling, under exhausting	
5. Soft swell	Advanced stage of swell at appearing both ends.	Over filling and under exhausting	
6. Hard swell	Final stage of swell. Can ends cannot be pressed back to concave position. Cans finally burst.	Over filling and under exhausting	
7. Panelling	Can body is pushed inside in large cans due to vacuum.	Thin tin plate, cooling at high pressures	Use cans prepared from good quality tin plate.
8. Leaker	Can contents leak out of the can	Defective seaming, nail holes created during putting nails on package, excess positive pressure due to spoilage inside the can, internal, external corrosion, improper handling and mechanical damage	Ensure proper exhausting, perfect seaming and adequate processing of the contents. Use properly lacquered cans.

9. Breathing	Tiny leakage allowing air to enter inside, inadequate vacuum. Food is damaged due to rusting.	Defective seaming, poor can quality	Use good quality cans and ensure proper seaming
10. Bursting	Bursting of can due to high pressure	Microbial spoilage or hydrogen swell	Use good quality cans, ensure proper, filling, exhausting, sealing and heat processing
11. Stack burning	Discolouration, cooked flavor, softening, pulpy product	Inadequate cooling of cans after heat processing before storage	Ensure proper cooling of cans before storage
12. Flat sour	Production of acid without formation of gas. Difficult to observed from outside	Spoilage due to *Bacillus sterothermophillus, B. coagulans*, under processing of cans.	Ensure proper processing so as to kill all the microorganisms.

Adapted from Srivastava and Kumar (2002)

Points to Remember

- √ Canning is the process of preservation of various food stuffs including fruits and vegetables whole or in pieces, in sugar syrup or brine by heat processing them in hermetically sealed containers.
- √ Nicholas Appert, a Persian confectioner is known as Father of Canning who invented the process of preserving foods in hermatically sealed glass containers by application of heat.
- √ Canning is also called as *Appertizing or Appertization.*
- √ Louis Pasteur proved in 1860s that microorganisms are the real cause of spoilage and by destroying them food can be preserved in suitable containers.
- √ Pasteurization refers to "preservation of food by heat treatment at sufficiently high temperature below 100° C to kill majority but not all of the microorganisms and by preventing their access to the food inside containers by sealing it hermatically".
- √ For canning fruits should be ripe but firm and vegetables should be tender
- √ Lye peeling is the process of removal of peel by in boiling caustic soda solution.
- √ Blanching is the treatment of fruits or vegetables with boiling water at about 100°C or steam for short periods of time followed by immediate cooling prior to canning.

- √ Blanching aims at inactivation of enzymes, while pasteurization aims at killing of most but not all of the microorganisms and sterilization aims at complete destruction of all of the microorganisms.
- √ Temperature for blanching, pasteurization & sterilization are 100, < 100 and > 100 (i.e.) 115-121° C respectively.
- √ Cooling is necessary after blanching to avoid over cooking, retain texture and integrity of fruits and vegetables and to give temperature shock to the microorganisms
- √ 100° Salometer reading = 26.5 % salt concentration (maximum solubility of salt in water is 26.5%)
- √ The temperature of syrup or brine at the time of can filling should be 79-82° C. Head space of 0.32 to 0.47 cm should also be kept at the time of filling syrup or brine.
- √ Partially seaming the lid to the can by a single first roller operation action of the double seamer is called Clinching. Lid remains sufficiently loose to permit the escape of free as well as dissolved air out of the can.
- √ Exhausting is done after clinching to remove practically all the air from the contents by passing the can through a tank of hot water at about 82-87° C for 5-25 min so that the temperatures at the centre of the can reaches not less than 79° C.
- √ Sealing is done after exhausting by the second roller operation of the double seamer. After sealing air can not enter into the can and cause any kind of spoilage.
- √ Heat processing is done after sealing of cans.
- √ Fruits can be processed satisfactorily at 100° C in boiling water as the presence of acids retards the growth of bacteria, while vegetables (except tomato) are required to be processed at 115-121° C.
- √ Cooling is necessary after heat processing to stop over cooking, prevent stack burning, give temperature shock to the microorganisms if any.

- √ The lids of can become concave if processed properly due to formation of vacuum inside.
- √ The trade size of A2½ cans is 401 x 411 and can contain 848 cm^3 of material
- √ Hydrogen swell occurs due to formation of hydrogen gas on reaction of metal and acids in food. Food generally is free from harmful microorganisms
- √ Stack burning occurs due to inadequate cooling of cans after heat processing before storage
- √ Flat sour is the spoilage of canned foods due to *Bacillus sterothermophillus, B. coagulans* and under processing of cans.

11

Drying and Dehydration

Drying/ Dehydration means the process of removal of water.

Drying : Removal of water (moisture) by using non conventional energy sources like Sun or wind.

Dehydration : Process of removal of moisture by the application of artificial heat under controlled conditions of temperature and / or humidity and air flow.

Principles of Preservation

- Hindering the growth and activity of microorganisms in absence of sufficient moisture
- Reduction of water activity. Microorganisms generally don't prefer to grow at water activity less than 0.50-0.65
- Non-availability/lesser availability of free water (eg in case of osmotic dehydration)

Factors Affecting the Rate of Drying

1. Composition of Raw Material

The food containing more of free water shall dry fast as compared to the food containing bound water.

2. Size, Shape, Arrangement of Food Particles

Larger the size, slower the drying and *vice versa*. Similarly, larger the surface area, i.e. thinner the slices, faster shall be the drying and *vice versa*.

3. Stacking of Produce

Stacking the produce in huge lots shall affect the rate of air movement through the inner portions of the food material kept for drying. In a huge stack the product may dry from outer side and the interiors may not dry as fast as the material kept towards the periphery.

4. Temperature, Humidity and Velocity of Air

Higher the temperature, faster the drying. But temperature beyond 60°C may affect the nutritional and sensory quality including colour, flavour etc. adversely. Generally drying is carried out at 50-60° C temperature.

Relative humidity of air. We generally observe that clothes do not dry easily after washing during rainy season, although the temperatures are quite high. On the contrary, temperatures are lower during winters but the clothes dry easily after washing. The reason behind this phenomenon is the relative humidity of air. During rainy season air is already saturated with moisture, so the moisture carrying capacity of air is very low. But the air is dry (less RH) during winters so the moisture carrying capacity is quite high. This is the reason why clothes dry faster during winters as compared to that during rainy season. The same facts shall hold true in case of dehydration of food. If the relative humidity of the air is high the food shall dry very slowly and shall dry quickly when the relative humidity is low.

Higher the air/ wind velocity, faster the drying and *vice versa* provided, the air is not saturated with moisture.

5. Pressure (barometric or under vacuum)

The boiling point of water is reduced at reduced pressures. Water may boil at 40-50° C at 28-30" Hg vacuum. Water also boils at lower temperatures as we go up on hills. Therefore, the rate of drying of food would also be faster at reduced pressure. Lesser the pressure faster the drying and *vice versa.*

6.Heat transfer to Surface (conductive, convective or radiative)

Conduction: The heat transfer takes places from one particle to the next particle without the movement of the particles in food eg. heating/ drying of solid food. This is relatively slower process and the outer particles may receive more heat as compared to the inner particles. The rate of heat transfer may depend on the conductivity of the food particles.

Convection : The heat transfer takes places with the movement of particles themselves. The food particles carry heat, create a density gradient, move from one place to other and may transfer heat to any particle which is not adjacent to the one carrying heat eg. heating/drying of liquid foods. This density gradient creates convection currents. This is relatively faster process.

Radiation : The heat transfer takes places by electromagnetic waves. eg heat transfer in an electric grill.

Types of Drying Processes

1. Sun Drying - Solar driers
2. Atmospheric Drying (air convention driers)

Kiln drier	Pieces
Cabinet, tray or pan drier	Pieces
Tunnel	Pieces
Continuous conveyer belt driers	Purees, liquids

	Belt trough driers	Small pieces, granules
	Fluidized bed dries	Small pieces, granules
	Spray driers	Liquids, purees
	Foam mat driers	Pulps and concentrates
	Drum/roller driers	Purees, liquids
3.	Sub-atmospheric dehydration	
	Vacuum shelf	Pieces, purees & liquids
	Vacuum belt	Purees and liquids
	Freeze Driers	Pieces, liquids

Psychrometry

It is study of inter relationships between temperature and humidity of the air.

Absolute humidity: Mass of water vapour/unit mass of dry air (kg/kg)

Relative humidity: Ratio of (partial pressure of water vapour in air) to (pressure of saturated water vapour at same temperature) x 100

Dry bulb temperature: Temperature of air measured by the bulb of thermometer

Wet bulb temperature: Temperature of air measured by the bulb of thermometer when it is covered by a wet cloth.

Note: Wet bulb temperature is always $\leq$ *dry bulb temperature.*

Hygroscopic and Non-hygroscopic Foods

Fruits and vegetables when dried generally have a strong tendency of absorption of moisture from the surrounding air and this property is called as *hygroscopicity* and such foods are called *hygroscopic foods*. In hygroscopic foods partial pressure of water vapour varies with the moisture content of food. Non-hygroscopic foods have a constant water vapour pressure at different moisture contents i.e. it does not vary with the moisture content of food.

Difference between Hygroscopic and Non-hygroscopic Food

Hygroscopic food	Non-hygroscopic food
• When kept in open air, they readily absorb moisture from the surrounding air	• When kept in open air, they do not readily absorb moisture from the surrounding air
• Partial pressure of water vapour varies with the moisture content of food	• Partial pressure of water vapour does not vary with the moisture content of food and remain constant at different moistures
• Spoilt easily in storage	• Not so
• Should be stored in air tight containers / packages with added anti-caking agents	• No need to add anti-caking agents

Some anti-caking agents are added for reduction of hygroscopicity of dried fruit and vegetable products. As per PFA table salt, onion powder, garlic powder, fruit powder may contain the following anti-caking agents in quantities not exceeding 2.0% either singly or in combination (PFA, 2008).

(i) Carbonates of calcium and magnesium

(ii) Phosphates of calcium and magnesium

(iii) Silicates of calcium, magnesium, aluminium or sodium or silicon dioxide

(iv) Myristates, palmitates or stearates of aluminium, ammonium, calcium, potassium or sodium.

Since most of the fruit products including beverages are prepared by addition of cane sugar in varying amounts, powdered sugar can also be successfully used as anti-caking agent in fruit powder i.e. lemon etc. (Sharma *et al.*, 2004, 2005).

Equivalent Relative Humidity (ERH) : It is the relative humidity at which the food will neither absorb or gain any moisture from the atmosphere nor it will loose any moisture to the surrounding atmosphere. Thus, ERH indicates the best relative humidity conditions for appropriate storage of a dried food.

$$\text{Water activity, } a_w = \frac{ERH}{100}$$

Sulphuring : The treatment where whole fruits, slices or pieces are exposed to the fumes of burning sulphur inside a closed chamber known as Sulphur Box for 30-60 min or in small air tight rooms in order to retain colour and prevent spoilage during drying and subsequent storage.

The sulphur dioxide fumes act as disinfectant and prevent the oxidation and darkening of fruits on exposure and thus improve their colour. The fumes also act as a preservative, check the growth of moulds etc and prevent the cut pieces from fermenting while drying in Sun.

Sulphitation - Dipping of fruits /vegetables pieces/slices in sulphur solution before drying/dehydration in order to retain colour and prevent spoilage during drying and subsequent storage.

Sweating - The process of stacking dried fruits in boxes/ bins to equalize the moisture contents before final packaging.

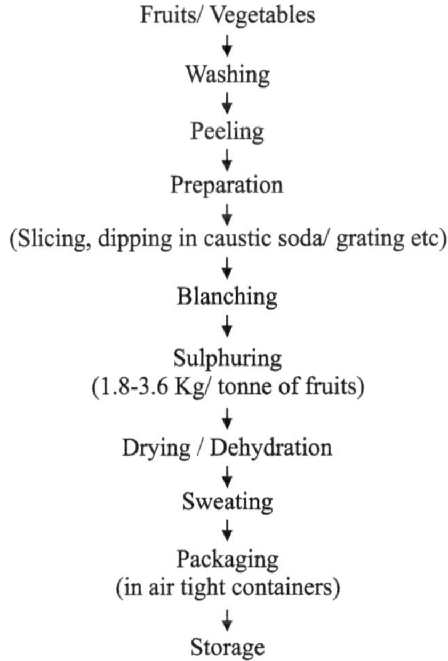

Drying/Dehydration of Fruits and Vegetables

Osmotic Dehydration

In this technique fruit pieces after lye peeling and/or blanching and sulphuring/sulphiting, are soaked in osmotic solution (60-70° Brix sugar solution) for some hrs to one - few days and taken out. Syrup is drained and product is dried in Sun or in a dehydrator. If the product is acidic, it can again be kept in the syrup after re-concentration to 60-70° Brix.

Three types of material transfer takes place

1. Sugar in
2. Water out
3. Minerals, acids, vitamins out (leaching into the syrup).

Principles of Preservation

High concentration of sugar (solute), low water activity below the level at which microorganisms can grow comfortably and partial final drying in air.

The rate of osmotic dehydration would primarily depend on

- Concentration of syrup - higher the concentration faster the moisture removal
- Temperature of syrup - higher the temperature faster the moisture removal. Too high temperatures beyond 50-60° C can affect the quality of the product adversely. Undesirable changes in colour, flavour and texture (softening) can take place.
- Frequency of agitation – frequent the agitation of material during osmotic dipping, faster would be the drying. Care should be taken not to damage the integrity of fruit pieces kept for dip.

Intermediate Moisture Foods (IMF)

These are products containing moderate levels of moisture (20-50%), and a low water activity (unsuitable for the growth of microorganisms) due to high concentration of dissolved solutes. IMF do not require refrigeration during storage and can be eaten without rehydration.

Characteristics of IMF

- Higher moisture content than dried foods
- Very low water activity
- High concentration of dissolved solutes
- Palatable, not crisp and not brittle
- Do not require refrigeration for storage.

Principles of Preservation

High concentration of dissolved solutes lowers down the water activity below the level at which microorganisms can grow comfortably. Osmotic pressure is very high and some preservative affect is also due to presence of sugar, salts, acids etc.

Examples of IMF

Honey, jam, jelly, cakes, dates, osmo-dried products.

Dehydrofreezing

Dehydration + freezing.

Moisture is removed partially by product dehydration before freezing.

Water Activity

Water activity (a_w) is the measure of effective concentration of water in a substance. It is the ratio of vapour pressure of water in the food to the vapour pressure of pure water at same temperature.

$$a_w = P_f/P_o = N_w/(N_w + N_s)$$

where P_f - vapour pressure of water in the food

P_o - vapour pressure of pure water
N_w - No. of moles of water
N_s - No. of moles of solute

Preferred water activity range by different microorganisms

Bacteria	0.86-1.0
Yeasts	0.78-1.0
Molds	0.65-1.0
No microorganisms	< 0.50

Difference between Conventional Drying and Freeze Drying

Conventional Drying	Freeze Drying
1. Temp- 37 to 93°C	1. Temp < 0°C
2. Pressure- usually atmospheric	2. Reduced pressure 4 mm Hg, 27-133 Pa
3. Short drying time usually < 12 hrs.	3. Drying time 12-24 hrs.
4. Principle-evaporation of water from food surface	4. Sublimation of ice to vapours with out passing through liquid phase
5. Odour frequently abnormal	5. Odour usually natural
6. Colour usually darker	6. Colour usually natural
7. Cost generally low	7. High cost
8. Reduced nutritional value	8. Nutritional value not changed
9. Structural damage and shrinkage after drying	9. Minimum structural changes and shrinkaege
10. Slow and incomplete rehydration	10. Rapid and complete rehydration

Difference between Sun Drying and Dehydration

Sun drying	Dehydration
1. Done in open sun light	1. Done in dehydrator under controlled conditions of temperature
2. Temperature is not under control	2. Temperature is under control
3. More time consuming	3. Fast process
4. Weather dependent, cannot be done under cloudy and rainy conditions	4. Does not depend on weather conditions
5. Chances of spoilage due to dust during windy conditions exist	5. No such dust spoilage occurs
6. Requires more floor space	6. Requires less floor space as many trays can be accommodated simultaneously in one lot
7. Inferior quality, colour	7. Superior quality, colour
8. More damage to nutritional and sensory quality	8. Less damage to nutritional quality
9. Product can not be dried to very low moisture contents	9. Product can be dried to very low moisture contents
10. Cost generally low	10. High cost

Difference between Sulphuring and Sulphiting/Sulphitation

Sulphuring	Sulphiting/ Sulphitation
1. Process of treating food pieces with sulphur fumes by burning sulphur in a closed chamber	1. Process of treating food pieces in sulphur solution before drying
2. Product remains dry even after treatment	2. Product takes some moisture from the solution
3. Sulphur chamber is required for treatment	3. Can be done in buckets of processing vessels

Points to Remember

- √ Drying/ Dehydration means the removal of water
- √ Drying is the removal of moisture under Sun or wind while Dehydration is the process of removal of moisture by the application of artificial heat under controlled conditions of temperature and / or humidity and air flow.
- √ Low moisture and low water activity are the main principles of preservation in drying and dehydration.
- √ Higher the temperature, faster the drying. Dehydration is generally done at 50-60° C temperature.
- √ Higher the air/ wind velocity, faster the drying and *vice versa* provided, the air is not saturated with moisture.
- √ Lesser the pressure faster the drying and *vice versa*.
- √ Conduction is the heat transfer that takes places from one particle to the next particle without the movement of the particles in food eg. heating / drying of solid food.
- √ Convection is the heat transfer that takes places with the movement of particles themselves.
- √ Radiation is the heat transfer that takes places by electromagnetic waves
- √ Spray drying is suitable only for liquid foods, beverages, milk and puree
- √ Psychrometry is the study of inter relationships between temperature and humidity of the air.

- √ *Absolute humidity:* Mass of water vapour / unit mass of dry air (kg/kg)
- √ *Relative humidity :* Ratio of (partial pressure of water vapour in air) to (pressure of saturated water vapour at same temperature) x 100
- √ *Dry bulb temperature:* temperature of air measured by the bulb of thermometer
- √ *Wet bulb temperature:* temperature of air measured by the bulb of thermometer when it is covered by a wet cloth. *Wet bulb temperature is always ≤ dry bulb temperature*
- √ Fruits and vegetables when dried generally have a strong tendency of absorption of moisture from the surrounding air and this property is called as *hygroscopicity* and such foods are called *hygroscopic foods*..
- √ The maximum permissible limit for the use of anti-caking agents is 2%.
- √ In *sulphuring* the whole fruits, slices or pieces are exposed to the fumes of burning sulphur inside a closed chamber known as Sulphur Box for 30-60 min or in small air tight rooms.
- √ In *sulphitation* fruits/vegetables pieces /slices are dipped in sulphur solution before drying / dehydration.
- √ *Sweating* is the process of stacking dried fruits in boxes/ bins to equalize the moisture contents before final packaging.
- √ *Osmotic drying/dehydration* is the partial removal of moisture from the fruit pieces by dipping in sugar solution followed by finishing in dehydrator. In osmotic dehydration sugar moves in, water moves out of the fruit pieces.
- √ Rate of osmotic dehydration would depend on concentration and temperature of syrup, and frequency of agitation.
- √ IMF contain moderate levels of moisture (20-50%), and a low water activity (unsuitable for the growth of microorganisms) due to high concentration of dissolved solutes.
- √ High concentration of dissolved solutes, low water activity and high osmotic pressure are the principles of preservation in IMF.

- √ Honey, jam, jelly, cakes, dates, osmo-dried products are examples of IMF.
- √ In dehydrofreezing moisture is removed partially by product dehydration before freezing.
- √ Water activity (a_w) is the measure of effective concentration of water in a substance. It is the ratio of vapour pressure of water in the food to the vapour pressure of pure water at same temperature.

12

Low Temperature Preservation

Freezing : Freezing is a process where temperature of food (fruits, vegetable etc.) is lowered down considerably below freezing point so that the water present inside the food undergoes change in state from liquid to solid (ice) and gets frozen immediately eg Frozen Pea. Water is immobilized and solutes are concentrated to reduce the water activity.

Principles of preservation : Combination of low temperature, low water activity, and pre-treatments like blanching, dipping in sugar syrup etc. stop (reduce) the growth and activity of microorganisms.

Theory of freezing : Freezing takes place in two main steps

(i) Sensible heat including heat of respiration in case of fresh fruits and vegetables is first removed to lower the temperature of the food to its freezing point.

(ii) Latent heat of crystallization is then removed and ice crystals are formed. Temperature is monitored at the thermal centre of food.

Frozen Storage refers to storage at temperatures generally -18° C or below. This will preserve perishable foods for months or even years. Below – 9.4° C there is no significant growth of microorganisms in the food.

Methods of Freezing

(i) *Sharp Freezing or Slow Freezing:* Temperature -15 to – 29° C and freezing may take 3-72 hrs. In this case ice crystals of large size are formed and they may also rupture and damage fruit cells.

(ii) *Quick Freezing:* Freezing time of 30 min or less and usually the freezing of small units or packages of foods. Temperature -17.8 to -45.6° C. The faster the freezing the smaller the size of ice crystals. In this case ice crystals of very small size are formed and they may not damage fruit cells.

Quick freezing can be accomplished by the following methods

a. *Direct immersion:* Food is immersed directly in sugar syrups or brines at low temperature, but the liquid must not freeze even at -18° C and should be edible. The liquids are good conductors of heat and make a perfect contact between fruit/ vegetable pieces and refrigerating medium, but the main problem encountered is that brine which has a very low freezing point cannot be used with fruits while sugar syrup becomes viscous and freezes at low temperature.

b. *Indirect contact with refrigerant:* Here food is in direct contact with a metal surface which is cooled by some suitable refrigerating media. So refrigerating media is not in direct contact with the food.

c. *Air blast freezing:* Cold air is circulated quickly through the food at -18 to -34°C.

(iii) *Cryogenic Freezing:* Ultra fast freezing at very low temperature of –60° C. Here the refrigerants used are liquid nitrogen and liquid carbon dioxide. Latent heat of vapourization is taken from the food.

Properties of Food Cryogens

Property	Liquid carbon dioxide	Liquid nitrogen	Dicholoro difluoro methane (refrigerant 12 or Freon 12)
Boiling point (°C)	-78.5	-196	-29.8
Consumption per 100 kg of product frozen (kg)	120-375	100-300	1-3

Source: Graham (1984)

(iv) *Dehydro-Freezing:* Half of the moisture of fruits and vegetable is removed by dehydration before freezing.

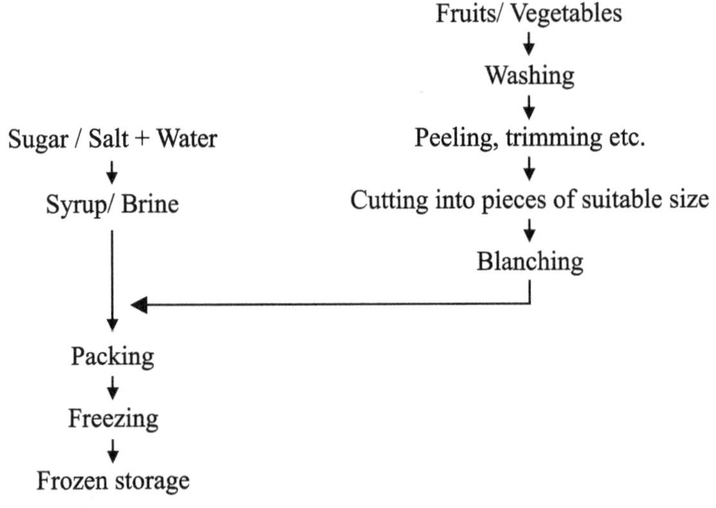

Freezing of Fruits and Vegetables

Thawing

Thawing means to pass from a frozen into a liquid or unfrozen state. When food is thawed in air or water, ice on the exterior surface of frozen food melts to form a layer of water. Since water has lower thermal conductivity and thermal diffusivity than ice, the melt ice (water) on the surface of thawed frozen food creates an insulating effect and the ice from the interiors does not melt very fast. During *thawing*, the surface layer of water reduce the rate of conduction of heat to the frozen interiors, as contrasted from that during *freezing* when formation of ice on the surface increases heat transfer (removal) from the interiors. Therefore, thawing is longer process than freezing even under similar temperature differences and other conditions.

Freeze Concentration : It is a method of water /moisture removal from the food without evaporation at freezing temperatures of 0°C or less, in which three distinctive steps are involved viz.

1. Ice crystallization or ice crystal formation done by freezing
2. Ice crystal enlargement in mixing vessel
3. Ice crystal separation from the concentrate phase by mechanical separation/filtration/sieving.

This is a very effective method of removal of water form food without any practical alteration of nutritional constituents including aroma volatiles. However, the process has high refrigeration costs and high capital costs of equipments. Degree of concentration is higher than that obtained in membrane concentration but lower than evaporative concentration.

Applications

Freeze concentration has its applications only in concentration of high value juices and extracts, vinegar etc. Freeze concentration is also used for pre-concentration of coffee extract prior to freeze drying. Also, it is used sometimes to increase the alcohol content of wines.

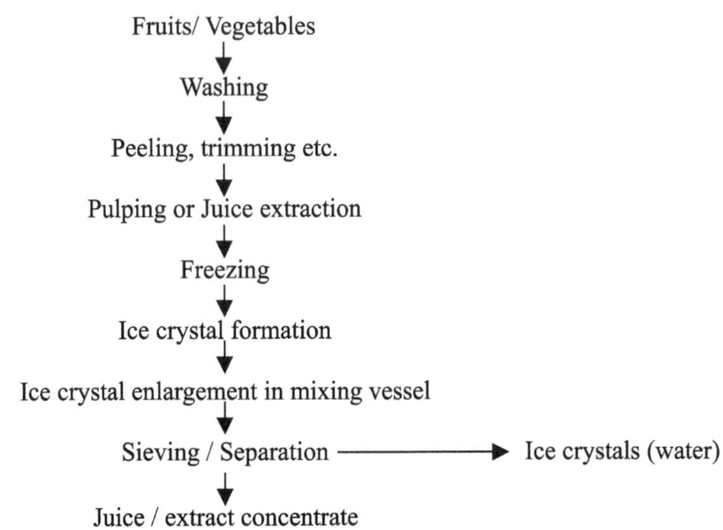

Freeze Concentration of Juices

Freeze Drying / Lyophilization : It is the process of removal of water by direct sublimation from frozen state (ice) to vapour state without passing through liquid water state using the triple point of water.

At *triple point* ice, water and vapours coexist in one state. Below triple point vapour pressure of ice is less than that of water which favours direct sublimation from ice to vapours.

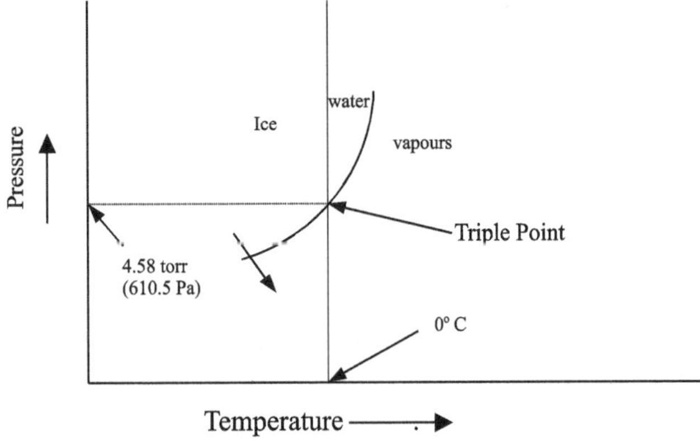

Fig. Sublimation of Ice at Triple Point

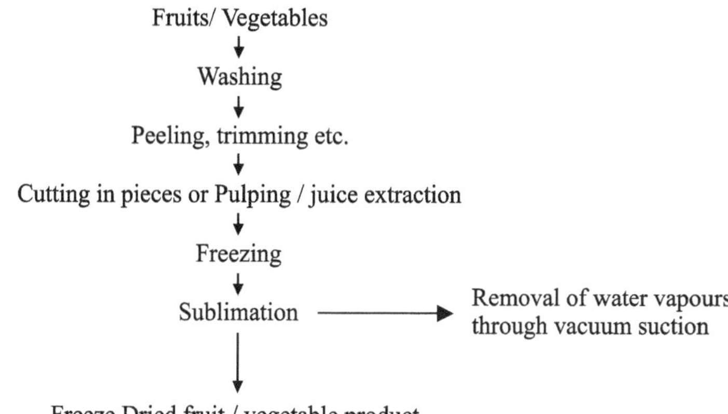

Freeze Drying of Fruits and Vegetables

Theory of freeze drying

(i) Food is frozen in conventional equipment

(ii) Water vapour pressure of food is held below 4.58 torr (610.5 Pa) and water is frozen

(iii) Food is heated by increasing the temperatures slightly and the ice formed due to food freezing (step i and ii) directly sublimes into vapours without entering into liquid state.

(iv) Water vapours are continuously removed through a vacuum pump and condensed on refrigeration coils.

Difference between Freezing, Freeze Concentration and Freeze Drying

Freezing	Freeze concentration	Freeze drying
1. Used in solid as well as liquid foods	1. Used in liquid food only	Used in solid as well as liquid foods
Water (moisture) is frozen only and is not removed out of the food	2. Water (moisture) is removed out of the food	2. Water (moisture) is removed out of the food
Phase change from liquid (water) to solid (ice) on freezing	3. Phase change from liquid (water) to solid (ice)	3. Phase change from liquid (water) to solid (ice) on freezing and gas (vapours) on sublimation
Done at atmospheric pressures	4. Done at atmospheric pressures	4. Done at reduced pressures

Points to Remember

- √ Freezing is a process where temperature of food (fruits, vegetable etc.) is lowered down considerably below freezing point so that the water present inside the food undergoes change in state from liquid to solid (ice) and gets frozen immediately
- √ Combination of low temperature, low water activity, and pre-treatments like blanching, dipping in sugar syrup etc. stop (reduce) the growth and activity of microorganisms are the principles of preservation involved in frozen products
- √ Frozen storage refers to storage at temperatures generally -18° C or below.
- √ Sharp freezing or slow freezing is done at -15 to – 29° C in 3-72 hrs.
- √ Quick freezing is done in 30 min or less at -17.8 to -45.6° C.
- √ Dehydro-freezing occurs when fruits and vegetable have about half their moisture removed by dehydration before freezing.
- √ Freezing can be done both in solid as well as liquid foods.
- √ Thawing means to pass from a frozen into a liquid or unfrozen state
- √ Freeze concentration is a method of water/moisture removal without evaporation and without any practical alteration of nutritional constituents including aroma volatiles.
- √ Freeze drying/lyophilization is the process of removal of water by direct sublimation from frozen state (ice) to vapour state without passing through liquid water state using the triple point of water.
- √ At *triple point* ice, water and vapours coexist in one state. Below triple point vapour pressure of ice is less than that of water which favours direct sublimation from ice to vapours.
- √ Freeze drying can be used in both solid and liquid foods while freeze concentration is used only in case of liquid foods.

13

Unfermented Fruit Beverages

Beverage: It includes all drinks whether they are fermented or unfermented, prepared from fruits or non-fruit, sweetened or unsweetened eg. tea, coffee, coca cola, cider, maaza are all beverages.

Fruit Beverages : Beverage which is prepared partly or fully from fruit juice or pulps with or without additives eg. pure fruit juice, squash, cordial, RTS beverage etc.

Non-fruit Beverage : Beverage which does not contain any fruit part eg. Coca cola, pepsi, sherbet etc.

Unfermented Juice or Pure Fruit Juice: This is the natural juice pressed out of the fruit and practically its composition remains unaltered during preparation and preservation. No additives like sugar, acid, water etc. are added. eg apple juice, orange juice.

Fruit Juice Beverage: This is the fruit juice that has been considerably altered in composition by adding water, sugar, acid, flavour, colours, preservatives etc. before consumption. It may or may not be diluted before it is served as a drink eg. squashes, nectars, appetizers etc.

Fermented Fruit Beverage: This is the fruit juice that has undergone alcoholic fermentation by *Saccharomyces cerevisiae* yeast. Fermented fruit beverages contain varying amounts of alcohol. eg. grape wine, apple cider, berry wine, plum wine, peach wine etc.

Fruit Squash: This is a product containing moderate quantities (min. 25% as per FPO) of fruit pulp or juice to which cane sugar is added for sweetening (final TSS > 40%) and sometimes acid is also added to make proper sugar acid blend eg. orange squash, lemon squash, pineapple squash, mango squash etc. The product is consumed after dilution in water.

Fruit Juice Cordial: This is sparkling clear sweetened fruit juice beverage similar to squash (containing equivalent fruit part but lesser TSS) from which all the pulp and other suspended material have been completely eliminated (generally by using clarification methods) eg. lime juice cordial. This also contains moderate quantities of fruit juice (min. 25%) and added sugar to attain a TSS of min. 30% in finished product.

Sherbet: This is a clear sugar syrup containing no fruit juice, pulp or extract and which has been artificially flavoured eg. Sandal sherbet, Kewra sherbet, Rose sherbet etc. Generally consumed chilled after dilution in water.

Syrup: This is a product prepared from clear syrup containing extract of fruits, herbs and flowers etc. eg. Brahmi Syrup, Rhodendron Syrup etc. It is consumed after dilution in water.

RTS Fruit beverage/ RTS Drink: It is a beverage which is consumed directly without dilution and contains min. 10% of fruit part as well as min. 10% TSS.

Nectar: It is a beverage which is consumed directly without dilution and contains min. 20% of fruit part and min.15% TSS.

Fruit Juice concentrate: Fruit juice which has been concentrated by the removal of water either by evaporation, freezing or membrane separation. i.e. evaporative concentration, freeze concentration or membrane concentration. Concentrates are diluted to single strength juices before conversion into carbonated beverages and other products eg. apple juice concentrate, pineapple juice concentrate etc.

Fruit Juice Powder: This is the fruit juice which has been converted into a free flowing, highly hygroscopic powder by dehydration or lyophilization. Natural fruit flavours in powdered form may be added to compensate for any loss of flavour during preparation of fruit juice powders. These are reconstituted readily to yield full strength, full fruit juice drinks e.g. Orange juice powder, lemon juice powder etc.

Preparation and Preservation of Fruit Juices

1. Selection and preparation of fruits

Freshly picked, sound fruits of suitable varieties at optimum stage of maturity should be used for juice extraction. Decayed or damaged fruits do not yield good quality juice and may impart undesirable taste and flavour to the juice or prepared beverage.

2. Washing

Fruits should be washed thoroughly with water to remove any adhering dust and extraneous matter. Residues of sprays of lead and arsenic should be removed with dilute HCl.

3. Juice Extraction

Generally juice from fresh fruits is extracted by crushing and pressing operations. During extraction, the juice should not be unduly exposed to air, as oxygen in the air will adversely affect the colour, taste and aroma and also reduce the vitamin contents of the juice.

4. Deaeration

The process of removal of dissolved oxygen along with other gases from the juice as immediately as possible after juice extraction is called *"Deaeration"*. The juice is subjected to high vacuum whereby most of the O_2 and other gases are removed.

5. Straining, Filtration and Centrifugation

The coarse suspended particles incorporated in the juice during the process of extraction are removed either by straining through sieves of non-corrodible metal or by sedimentation. These suspended particles primarily include broken fruit tissues, seeds and skin besides the presence of colloidal suspensions of gums, pectic substances and proteins etc.

Methods of Juice Clarification

a. *By Sedimentation:* Juice is stored in carboys or barrels, adding a chemical preservative to prevent fermentation of juice during storage. The coarse particles settle down gradually within few days. The supernatant juice which contains mostly fine suspended particles, and colloidal suspensions, is siphoned off for further use.

b. *By use of fining agents:* This method produces a voluminous flocculant precipitate which settles gradually at bottom of the container, carrying down with it finely divided particles and colloidal suspensions that are responsible for cloudiness or haziness in the clarified juice.

Fining agents are of 3 types

i. *Enzymes:* These destroy colloids present in the juice. Various kinds of enzymes such as proteolytic, pectin-decomposing, starch liquefying enzymes are used eg. Proteases, Pectinol, Pectinase CCM, Filtragol, Amylase etc.

ii. *Fining agents having purely physical or mechanical action:* Commonly infusorial earth such as Spanish Clay, Kaoline, Bentonite, China Clay etc. are used commercially for mechanical fining or clarification. Concentration to be used is about 0.5-1.0 %. *Absorbing Carbons* are not suitable for clarification of fruit juices, because they absorb colouring matter of juice, remove tannins and flavouring compounds thereby rendering the juice insipid.

iii. *Chemical fining agents:* The colloidal substances in fruit juices carry a negative charge, and are precipitated when the charge is neutralized. Gelatin and Casein act partly in this manner and sometimes by forming insoluble precipitates with the juice constituents eg. Casein combines with the organic acids and gelatin combines with the tannins.

c. *Clarification by Freezing:* Colloidal suspensions are readily precipitated on freezing. Freshly extracted grape juice contains suspension of pulp, skin etc. and varying quantities of *cream of tartar* or *Potassium Hydrogen Tartarate*. When the juice is bottled, the cream of tartar precipitates gradually in the form of fine crystals which is undersirable. In order to avoid crystal formation due to slow precipitation, the bulk of the juice is subjected to refrigeration for several months to complete the precipitation. The clear juice is then bottled. This process is commonly observed in grape juice and wine so both are treated similarly.

d. *Clarification by heating:* Colloidal material in fruit juices usually coagulates when the juice is heated and settles down readily. Commonly, the juice is heated to about 82° C for 1-2 min. and then cooled rapidly. Pomegranate juice is clarified by this method. Other juices i.e. lime, lemon juice etc. also respond well to this treatment.

Similarities and Differences between Fruit Juice and Pulp

Pulp	Fruit Juice
1. Both are used as raw material for the preparation of other finished beverages and products	
2. Methods of preservation are also similar for both	
a. Pulp is thicker in consistency	a. Less thick in consistency
b. Contains more of suspended and insoluble solids	b. Lesser suspended solids and more dissolved solids
c. Extracted through pulper	c. Extracted through juice extractors, crushing and pressing operation
d. Pulp is mostly extracted from fruits/fruit pieces after heating/ cooking for few minutes.	d. Generally fruits are not heated/ cooked before juice extraction
e. eg mango pulp, apple pulp, plum pulp, guava pulp etc.	e. eg apple juice, orange juice, lemon juice, tomato juice

6. Preservation of Fruit Juices and Pulps

Fruits juice after extraction may be spoilt quickly due to fermentation by microorganisms, colour and flavour changes by enzymes, taste and aroma changes by chemical reactions, oxidative changes etc. In order to avoid these undesirable changes in juice quality and to retain the natural taste and aroma, it is necessary to preserve it soon after extraction. The methods of preservation generally used are:

A. Pasteurization

Heating of juice or pulp at high temperatures below 100°C for a sufficiently long time to kill most but not all of the microorganisms present in the food, followed by immediate cooling to room temperature. Some of the spores and spore forming bacteria i.e. *Bacillus subtilis* and *Bacillus mesentroides* can survive the process.

There are two ways of application of heat to the juice or pulp.

a. *LTLT:* Low Temperature Long Time

b. *HTST:* High Temperature Short Time (better than LTLT due to better retention of nutritional quality i.e. vitamins).

Flash Pasteurization: A process in which the fruit juice is heated rapidly for a short time of about 1 min. at a temperature about 5.5 °C higher than the normal pasteurization temperature for the juice. The juice after treatment is then filled into containers which are sealed hermetically under the cover of steam to sterilize the seal and then cooled. This process was developed especially for canning of natural orange juice but can be adopted for pasteurization of other fruit juices like apple, grape, lime, lemon juice etc.

Advantages of flash pasteurization:

a. It minimizes loss of flavour volatiles
b. Helps in retention of vitamins and other nutrients
c. Keeps the juice uniformly cloudy
d. Minimizes any cooked taste or burning flavour
e. Effects the economy in time and space

UHT Processing : Treatment of food at extremely high temperatures 130-150° C for extremely short durations of time (2-3 sec). The rate of destruction of microorganisms is very fast at high temperatures as compared to the rate of loss of nutritional and functional value of food. By the time chemical reaction starts the microorganisms are killed and treatment is stopped so there is practically little change in nutritional quality.

Pasteurization Treatment in Milk

LTLT	-	62.8° C for 30 min.
HTST	-	71.7° C for 15 sec.
UHT	-	137.8° C for 2 sec.

B. Preservation with Chemicals

Preservative : British Food and Drug Act, 1928 defines preservative as "any substance which is capable of inhibiting, retarding or arresting the process of fermentation, acidification or other decomposition of food or of masking any of the evidences of any such process or of neutralizing the acid generated by any such process, but does not include common salt, sugar, salt-peter (sodium or potassium nitrate), acetic acid/ vinegar, alcohols or potable spirits, spices, essential oils or any other substance added to the food by the process of curing known as smoking".

So *preservative* can be defined as a chemical additive having prescribed limits of use as per law, added delebrately to the food for prevention or delay of spoilage.

Commonly used Preservatives are

1. *Sodium Benzoate:* This is the salt of benzoic acid. Benzoic acid is sparingly soluble in water but its sodium salt is about 170 times more soluble in water. Sodium Benzoate is practically tasteless and odourless. Benzoic acid is more effective against yeasts than mould and its preservative action increases in the presence of CO_2.

   Mode of action : It affects the permeability of cell membrane of the microorganisms. The cellular fluids come out and the microorganisms die.

2. *Sulphur Dioxide* (K_2O2SO_2 or $K_2S_2O_5$ Potassium Metabisulphite or KMS)

   KMS is a dry chemical and is easier to use than liquid or gaseous Sulphur Dioxide.

   Mode of Action

   a. Breaks the S-S bonds in cysteine, methionine (S containing amino acids) and cause denaturation of proteins so that they are not available to the microorganisms. The process is called "Sulphytolysis". Some microorganisms themselves are proteins eg. yeasts, so are killed.

   b. Supply of carbohydrates to the microorganisms is checked.

c. Vitamins (Vit B_1 / Thiamine)

   Vit B_1 + SO_2 $\rightarrow$ Compound formed is unavailable to the microorganisms.

d. SO_2 reduces O_2 tension in the food and cause anaerobiosis.

3. *Parabens (Para hydroxy benzoic acids):*

   These are more or less independent of pH.

   Mode of Action:

   a. They attack protein to form phenolic group

   b. They interfere with folic acid metabolism which is a microbial growth factor and is an analogue to parabens so stops the energy processes.

      Types of Parabens: Butyl, Ethyl, Methyl, Propyl Parabens.

4. *Nitrates*

   These are the important preservatives for meat products and used to fix red colour in meat and its products.

   Mode of Action

   a. They affect genes and cause change in the base pairs and prevent replication

   Adenine $\xrightarrow{KNO_2}$ Uracil $\rightarrow$ No replication

   b. Cause mutational changes in the genes and effect genetic reconstitution

   - *Types of nitrates used* : KNO_2, KNO_3, $NaNO_2$, $NaNO_3$.
   - The excessive use of nitrates form nitrosamines which can lead to cancer.

5. *Sorbates*

   These are the salts of sorbic acid and used in bakery products as anti molds. Sorbic acid is 2,4 hexa dienoic acid. $CH_3 - CH = CH - CH = CH - COOH$

Mode of Action

a. It affects the enzyme activity (Succinic dehydrogenase and fumarase enzymes)

   It is the safest of all preservatives & is easily excreted from the body. It enters β oxidation and forms pyruvic acid.

C. Preservation with Sugar

Fruit products containing 66% or more sugar do not generally ferment. Food products containing > 66% sugar (specific gravity 1.330) have little free moisture available for microorganisms so their multiplication is inhibited and they die gradually.

Mode of Action

a. It binds water and reduces availability of free water and thus act as humactant (compounds that bind water)
b. Reduces water activity
c. Increase osmotic pressure.

D. Preservation with Salt

Salt acts as preservative above 12 % concentration.

Mode of Action

a. It binds water so that it is not available to micro-organisms.
b. Reduces water activity
c. Increase permeability of microbial cell membrane so only haloduric (salt tolerant microorganisms) can grow.
d. Cl^- ions are oxidizing agents therefore are toxic and bactericidal
e. Salt effects the solubility of O_2 so aerobic micro-organisms die.
f. Salt effects the enzyme system particularly proteases so protein synthesis stops and microorganisms will die.

E. Vegetable / Mustard Oils

Mode of Action

a. Creates hydrophobic regions

b. Contain an antimicrobial compound called "Isothio cyanate"

Anti microbial compound in clove are "Eugenol and Allicin".

Difference between Class I and Class II Preservatives

Class I preservatives	Class II preservatives
1. Do not have any maximum permissible limits of their use	1. Maximum permissible limits have been prescribed in FPO and PFA
2. eg sugar, salt, oil	2. eg Potassium metabisulphite, Sodium benzoate, sorbates
3. Not harmful if consumed in excess	3. Very harmful if consumed in excess

Types of Fruit Beverages

The beverages are generally prepared from the preserved and stored juices and pulps. In case the beverage is to be prepared from fresh juice or pulp, the extraction and preservation of juices may be done as detailed earlier in this chapter. Nectar is always prepared from fresh juice and should not contain any preservatives or artificial colours or essences.

1. Fruit Squash

Fruit squash is a beverage which essentially consists of moderate quantity of fruit pulp/juice (min. 25%) to which cane sugar is added for sweetening to raise TSS above 40°Brix and is consumed after dilution preferably with chilled water. eg. orange squash, plum squash etc.

- Squash should have min. TSS of 40% in final product and fruit juice or fruit part should be min. 25%.

- Juice or pulp is extracted from different fruits and is pasteurized.
- Sugar and citric acid (if needed), flavouring material, colours and preservatives are added in correct proportion.
- Sugar, citric acid and water are mixed and heated.
- The dirt is skimmed off. Clean syrup is blended with juice. After mixing all the ingredients a calculated amount of Sodium Benzoate or KMS is added.

  SO_2 not more than 350 ppm

  Sodium Benzoate not more than 600 ppm
- Bottles should be cleaned and sterilized / disinfected by dipping in 1 % KMS solution before filling.
- About 1.2 to 2.5 cm head space should be kept.

Steps

a. Filter the juice through muslin cloth and homogenize it thoroughly if required.

b. Calculate the quantities of additives required for preparation of a known quantity of squash as per FPO specifications.

c. Add weighed sugar and acid (if any) to weighed quantities of water to prepare the sugar syrup by boiling. Remove the scum, filter the syrup and cool.

d. Mix the weighed amount of fruit juice in the syrup to dissolve thoroughly.

e. Add weighed amount of preservative, colour or essence by taking out small quantity of prepared squash in a beaker, dissolve them and adding finally to the whole lot

f. Fill in squash bottles. Seal with PP Cap sealing machine and label before storage.

Squash should have TSS of about 40-45 % and acidity of about 1.0-1.2 % so that the final product obtained after dilution has a TSS of about 10-12 % and acidity of 0.2-0.30 %. The calculations for preparation of all the squashes shall depend on the initial TSS and acidity of the raw material i.e. pulp or juice.

Recipes

Fruit	Juice/ Pulp (kg)	Sugar (Kg)	Water (L)	Citric acid (g)	KMS (g)	Sod. Benzoate (g)	Colour / Essence
Apricot	1	1.8	1.2	25	1.5	Not added	-
Plum	1	2	1	-	Not added	3	-
Mango	1	1.5-2.0	1.2	25-30	2	Not added	-
Litchi	1	1.25	0.75	30	1.8	Not added	-
Lime/ Lemon	1	1.8	1.2	-	3	Not added	Optional
Orange	1	1.8	1.2	25-30	3	Not added	Optional

Source: Vaidya and Vaidya (2000)

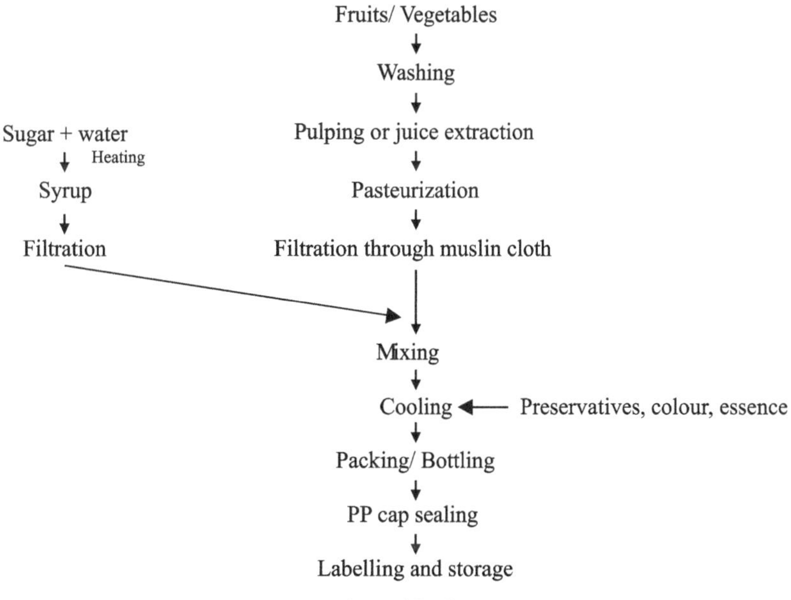

Preparation of fruit squash

2. Fruit Appetizer

Fruit appetizer is a beverage (also called spiced squash) which essentially consists of moderate quantity of fruit pulp / juice (min. 25%) to which cane sugar is added for sweetening to raise TSS above 40°Brix alongwith added salt and spice extracts and is consumed after dilution preferably with chilled water as is done in case of squash. eg. plum appetizer, apricot appetizer.

- The appetizing affect of these beverages is due to the presence of spices, condiments and herb extracts. Since appetizers are similar to squashes except for the presence of spices and condiments they are also called "Spiced Squashes"
- The extract of mentha is prepared by grinding properly washed leaves and extraction of juice.
- All spices after converting into powder are boiled with sugar syrup and strained. The rest of the procedure and specifications are same as that for preparation of squash.

Plum / Apricot Appetizer

Plum/ Apricot Pulp	1 Kg
Sugar	1.6-1.8 Kg
Water	500 -600 ml
Mentha extract	20 ml
Salt (White + black)	25 + 15 g
Citric acid	5-6 g
Dry ginger	5 g
Cardamom large	5 g
Cumin	10 g
Black Pepper	8-10 g
KMS	1.5 g (apricot), Sodium benzoate (1.5-2 g in plum appetizer).

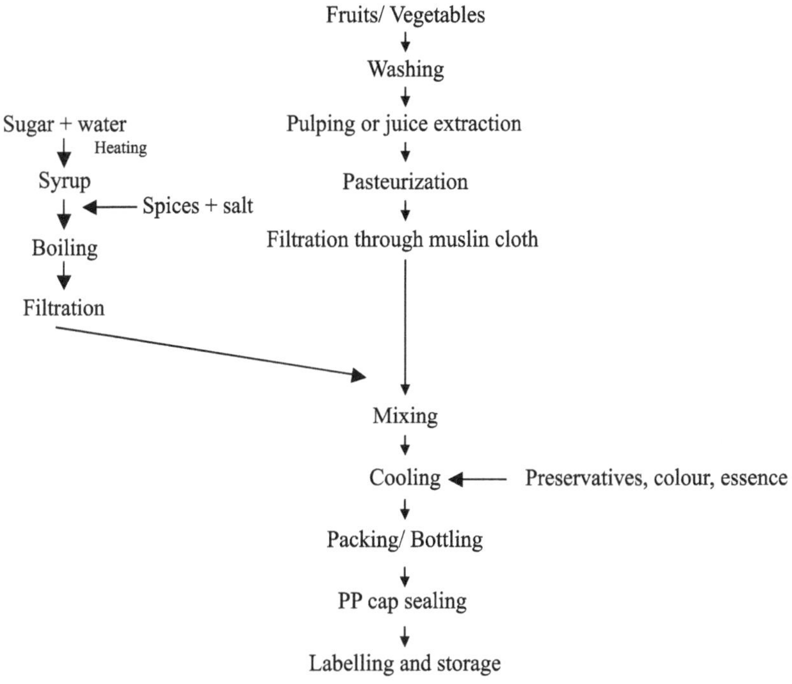

Preparation of Fruit Appetizer

3. Fruit Juice Cordial

It is sparkling clear, sweetened (TSS > 30%) fruit juice beverage similar to squash containing moderate quantities of fruit juice (> 25%) from which all the pulp and other suspended material have been completely eliminated by using juice clarification methods. The juice is clarified till it is sparkling clear by using various methods of juice clarification and the rest of the procedure is the same as that for preparation of squash.

Specifications

TSS not less than 30 %

Fruit part not less than 25 %

Acidity not more than 3.5 %

Preparation of Lime Juice Cordial

Cordial is a product that should essentially be prepared from a clarified juice so that the finished product does not have any suspended insoluble material. Therefore the clarification of lime juice is must before starting the preparation of product.

Clarification of Juice

Various clarifying agents i.e. gelatin, egg albumin, casein, mixture of gelatin and tannin are used for clarification of juices. These clarifying agents are added to the juice alongwith KMS and stored in glass carboys or in upright wooden barrels for about 2-3 months to get a clear juice, which is decanted or siphoned for use. This process is slow and time consuming.

Enzymes, such as pectinase and cellulase are used commercially for the clarification of juice. The juices should be treated with these enzymes @ 0.2 % for about 2 hrs at 50 ± 2°C followed by filtration under suction through filter press to get a sparkling clear juice in a short time.

Steps

1. Calculate the quantities of clarified juice, sugar and water required for preparation of a known quantity of lime juice cordial as per FPO specifications.
2. Prepare syrup of 40°Brix. Cool and filter through thick muslin cloth.
3. Mix the clarified juice and filtered syrup and add edible colour, preservatives etc.
4. Fill product in pre-sterilized glass bottles leaving about 2.5–3.0 cm head space and are sealed.
5. Dry and label the bottles.

4. RTS Beverage

It is a beverage which is consumed directly without dilution and contains minimum 10% of fruit part as well as minimum 10% TSS.

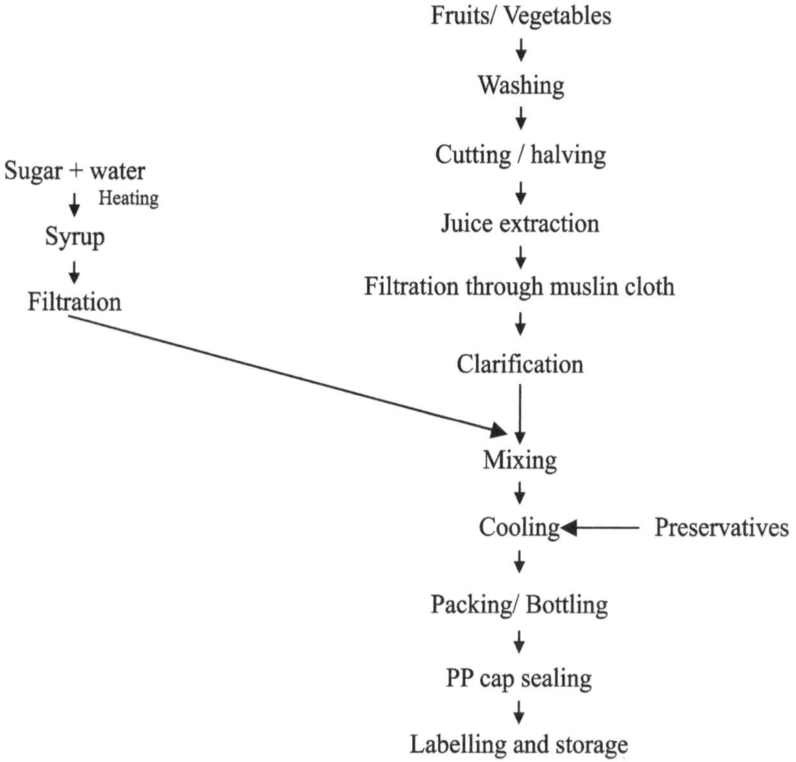

Preparation of Lime Juice Cordial

Steps

a. Filter the juice through muslin cloth to remove any suspended material.

b. Calculate the quantities of juice, sugar, water and acid (if required) required for preparation of a known quantity of RTS beverage as per FPO specifications.

c. Add calculated amount of sugar and acid to weighed amount of water and heat to boiling and filter through muslin cloth.

d. Add the weighed amount of juice to the boiling mixture. Heat to about 85-87°C or practical boiling.

e. Add the calculated amount of essence and colour dissolved in small amounts of the mixture (step d) towards the end of heating.

f. Hot fill the beverage into pre-sterilized glass bottles upto the brim and cork seal them immediately.

g. Sterilize the sealed bottles in boiling water for about 10-15 min and cool rapidly to room temperature.

h. Dry and label the bottles.

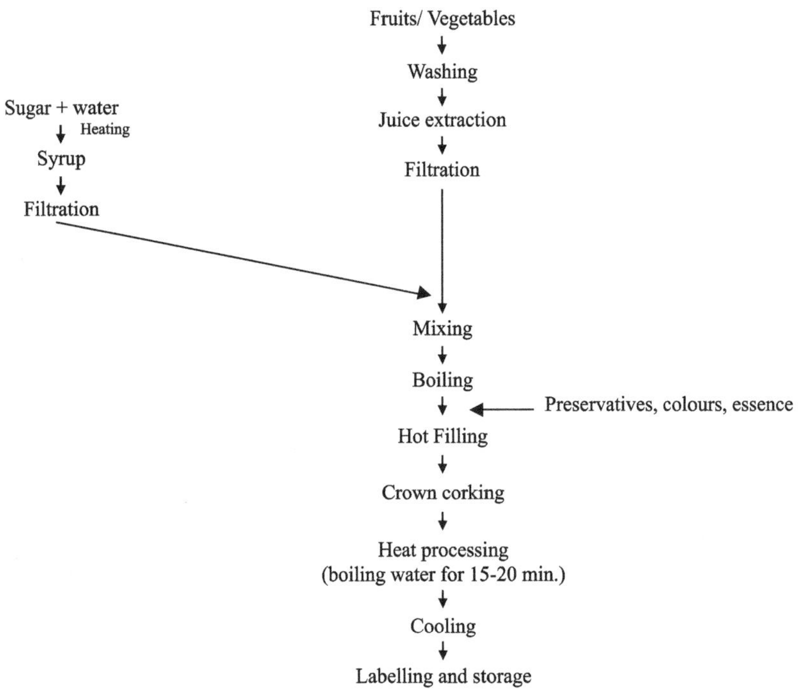

Preparation of RTS Beverage

5. Nectar

It is a beverage which is consumed directly without dilution and contains min. 20% of fruit part and min.15% TSS. Nectar is always prepared from fresh juice and should not contain any preservatives or artificial colours or essences.

Steps

a. Extract the juice as discussed earlier in this chapter.

b. Filter the juice through muslin cloth to remove any suspended material.

c. Calculate the quantities of juice, sugar, water and citric acid (if required) required for preparation of a known quantity of nectar as per FPO specifications.

d. Gradually mix the weighed quantities of water into the pulp/juice.

e. Add calculated amount of sugar and acid and mix thoroughly.

f. Strain the product through muslin cloth and heat to 85-87 °C or practical boiling.

g. Hot fill the beverage into pre-sterilized glass bottles upto the brim and cork seal them immediately.

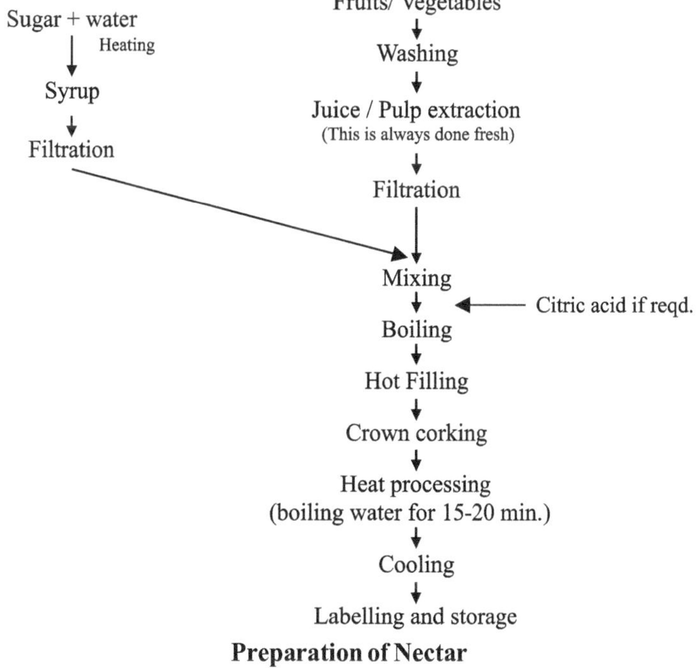

Preparation of Nectar

h. Heat the sealed bottles in boiling water for about 10-15 min and cool rapidly to room temperature.

i. Dry and label the bottles.

Difference between Squash and Cordial

Squash	Cordial
• Contains min. 40% TSS	• Contains min. 30% TSS
• Not a sparkling clear product	• Sparkling clear product devoid of any suspended material
• Prepared from many fruits i.e. mango, pineapple, litchi, lemon, orange, plum etc	• Mostly prepared from lime juice

Difference between RTS Beverage and Nectar

RTS beverage	Nectar
• Contains min. 10% TSS	• Contains min. 15% TSS
• Contains min. 10% fruit part	• Contains min. 20% fruit part
• Artificial colours and essence can be added	• Artificial colours and essence can not be added
• Can be prepared from fresh/ preserved pulp/ juice	• Generally prepared from fresh pulp/ juice

6. Carbonated Beverages

Carbonation : It is the process of dissolving or incorporating carbon dioxide into a ready to serve fruit or non-fruit beverage so that when served it gives off gas in fine bubbles and has a characteristic pungent taste. Such beverage containing added (not self generated) dissolved carbon dioxide under pressure is called as carbonated beverage. Coca-cola, Pepsi, Fanta, Sprite etc. are common examples of non-fruit carbonated beverages.

- Clarification of fruit juices is essential for preparation of carbonated drinks.
- For longer storage, the use of preservatives is important.

Lime Carbonated Drink

Juice	3%		
Acidity	0.2 %		
TSS	10-12° B		
CO_2	100 psi		
or	Brix acid ratio	=	50
	CO_2	=	100 psi

7. Juice Concentrate

Juice concentration is a process of separation and removal of water from the solids of the juice by making use of the basic three methods i.e. evaporation, freezing and membrane technology. Pulps and juices are the basic material used for preparation of various types of processed products. However, preservation of single strength juice is not at all economical due to the presence of large quantities (75-90%) of water in different juices. Concentration of juices not only provides microbial stability, but also reduce the costs of packaging, transportation and storage of juices by reducing the bulk of the material.

Methods of Concentration

1. *Evaporative concentration:* This method involves the removal of water from the juice by converting the liquid into vapours at low temperature (about 50° C) under vacuum in order to retain maximum nutritive and sensory quality. Here we use the difference in boiling point of solute and solvent as the basis for separation. This method is not very expensive and is practically used at commercial levels for the preparation of concentrates. Some loss to the nutritional and sensory quality of juice occurs during the process of concentration but we can achieve very high TSS (> 70°Brix) of concentrates by this method.
2. *Freeze concentration:* This process involves the separation of water from the juice by freezing the liquid into ice and

then separation of ice crystals. This process uses the difference in freezing point of solute and solvent as the basis for the separation. The equipment is very costly and loss of juice solids may occur due to entrapment in ice crystals. Quality of concentrate is superior than that obtained by first method. We can not achieve very high TSS of concentrate by this method.

3. *Membrane concentration:* This process involves the separation of water from the solids by passing the juice under pressure through a membrane which permits only smaller particles to pass through it. Bigger sized particles are retained on the top. It uses the difference of particle size as the basis of separation. Quality of concentrate is superior than that obtained by first method. We can not achieve very high TSS of concentrate by this method.

Clarification of juices: For achieving high TSS of the concentrates clarification of juices by enzymatic methods is very important. Unclarified juice when concentrated may lead to formation gel at high concentrations and will not be free flowing. Clarification is generally done by using pectinase, cellulase etc. enzymes @ 0.2-0.3 % at 50 ± 2°C for 2 hrs followed by filtration under suction / pressure through filter press.

8. Crush

It is a beverage similar to squash which is consumed after dilution in water and contains min. 25% of fruit part and min. 55% TSS. The acidity is kept at about 1.0 - 1.2%.

9. Syrup

It is a beverage similar to squash which is consumed after dilution in water and contains min. 25% of fruit part (pulps / juice or most commonly extracts of rose, sandal, brahmi etc.) and min. 65% TSS. The acidity is kept at about 1.3%.

10. Barley Water

It is a beverage similar to squash which is consumed after dilution in water and contains min. 25% of fruit juice (lime, lemon, grapefruit, orange etc.) and min. 30% TSS alongwith 0.25% barley strach. The acidity is kept at 1.0 %.

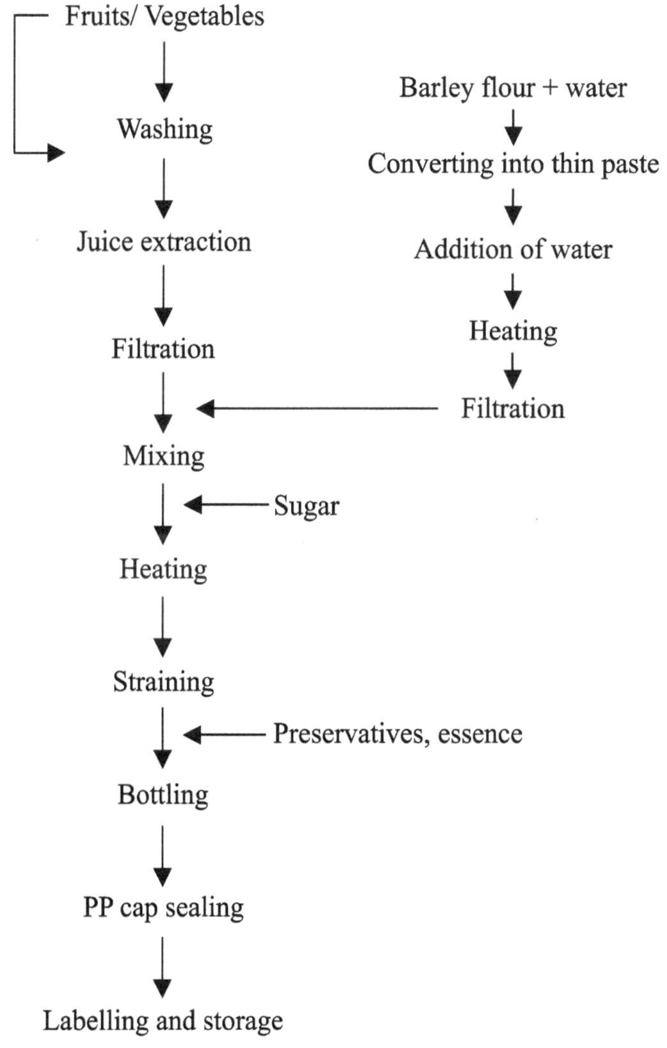

Preparation of Barley Water

Points to Remember

√ Tea, coffee, coca cola, cider, maaza are all beverages.

√ No additives like sugar and acid, water etc. are added in pure fruit juice

√ The composition of fruit juice is altered by addition of sugar, salt, acid, colour, preservatives, essence etc. for preparation of different beverages.

√ Wine and cider are fermented beverages while RTS and squash are unfermented beverages.

√ Fruit squash and cordial consist of min. 25% fruit part and final TSS > 40% and >30% respectively.

√ Sherbet contains no fruit juice, pulp or extract while syrup is prepared from extract/ juice.

√ RTS drink contains min. 10% fruit part and 10% TSS while nectar contains min. 20% fruit part and 15% TSS.

√ Concentration can be done by evaporation, freeze concentration and membrane technology.

√ Fruit juice concentrate should have a min. TSS of 32° Brix and 100% fruit part.

√ The process of removal of dissolved oxygen along with other gases from the juice immediately after juice extraction is called *"Deaeration"*.

√ Heating of juice to less than 100°C for a sufficiently long time aiming to kill most but not all of the microorganisms present in the food, followed by immediate cooling to room temperature is called as Pasteurization.

√ Flash Pasteurization is a HTST process in which fruit juice is heated only for a short time at a temperature of about 5.5°C higher than the normal pasteurization temperature for the juice. This reduces loss of nutritional quality and flavor.

√ Treatment of food at extremely high temperatures 130-150° C for extremely short durations of time (2-3 sec) is called as UHT processing.

- √ Sodium benzoate is used in coloured juices and is 170 times more soluble in water than benzoic acid
- √ Parabens are Para hydroxy benzoic acids and are more or less independent of pH.
- √ Nitrates are used as preservatives in meat products to fix red colour.
- √ Sorbates are salts of sorbic acid and are more safer than other preservatives.
- √ Sugar acts as preservative above 66% while salt above 12% concentration.
- √ Nectar, RTS drink and carbonated beverages are consumed directly without dilution while squash, appetizer, cordial, syrup, crush, barley water, concentrate are all consumed after dilution in water.

14

Alcoholic Beverages and Vinegar

Alcoholic Beverages

These are the beverages which are prepared after alcoholic fermentation of sugars by yeast, contain varying amounts of ethyl alcohol (5-42%), and are consumed directly or after dilution in water.

Wine : Product made by alcoholic fermentation of grapes or grape juice unless otherwise specified, by yeast (*Saccharomyces cerevisiae* and a subsequent ageing process. Alcohol content is 11-14 %, but may be as low as 7%.

Difference between Still Wines and Sparkling Wines

Still wines	Sparkling wines
• Retain no CO_2 • eg Cider	• Contain considerable amount of CO_2 • eg Champagne

Difference between Dry Wines and Sweet Wines

Dry wines	Sweet wines
• Contain no or little unfermented sugars	• Contain unfermented sugars or is added later on

- *Fortified wines:* Contain added alcohol / distillate of wine (brandy). Alcohol % is 19-21 %.
- *Table wines:* low alcohol and little or no sugar
- *Dessert wines:* Fortified sweet wines.

Sherry	Produced by special technique (secondary fermentation by special type of "Flor yeast" or by baking), containing 18-21% alcohol, could be sweet or dry.
Cider	Apple wine.
Soft Cider	1-5 % alcohol
Hard Cider	5-8 % alcohol
Apple wine	> 8% alcohol but can be as high as 14%.
Perry	Pear wine (could be sweet or dry)
Mead	Honey wine
Boukha	Fig wine
Toddy	Coconut wine
Vodka	Prepared from potato
Fanny	Cashew apple wine
Tokay	Famous wine made in Hungary
Port wines	Made from grape must fortified with brandy. Produced in Douro region of Portugal (alcohol about 18%).
Nira	Prepared from juice of palm tree
Berry wine	Prepared from strawberry, blackberry etc.
Sake	Yellow rice beer of Japan (14 to 17% alcohol)
Sonti	Rice beer of India (mould *Rhizopus sonti* is used)
Pulque	Latin American fermented beverage prepared from juice of agave or century plant (6% alcohol)

Vermouth	Fortified wine with 15-21 % alcohol, flavoured with a characteristic mixture of herbs and spices, some of which impart an aromatic flavour and odour while others a bitter flavour (may be sweet or dry).
Brandy	It is an alcoholic distillate from fermented grape juice or wine unless otherwise specified.
Rum	Distillate from alcoholically fermented sugarcane juice or molasses
Whiskey	Distillate from fermented grains, rye, wheat.
Beer	Prepared by fermentation of barley grains

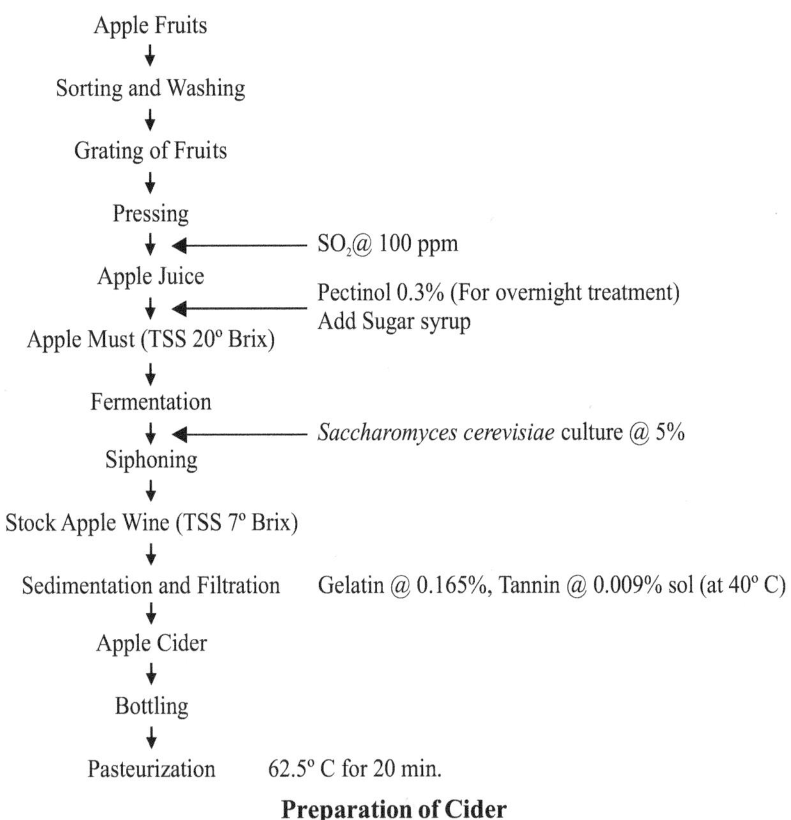

Preparation of Cider

Difference between Light, Medium and Strong Wines

Light wines	Medium wines	Strong wines
• Alcohol 7-9% • Not fortified with brandy	• Alcohol 9-16% • Sometimes fortified with brandy	• Alcohol 16-21% • Are definitely fortified wines

Difference between Red Wines and White Wines

Red wines	White wines
• Prepared from red/ blue coloured grapes have high anthocynin content • Fruit skin is left in the must for colour extraction	• Prepared from greenish grapes contain less anthocynin content • Skin is sometimes removed

Vinegar

Vinegar is a condiment prepared by alcoholic fermentation of carbohydrates of the starch/ sugar containing commodities i.e. fruits, potato, molasses etc. or followed by acetous fermentation.

Types of vinegars: 2 main types i.e. brewed and artificial vinegar.

Difference between Brewed Vinegar and Artificial Vinegar

Brewed vinegar	Artificial / synthetic/ non brewed vinegar
• Prepared from natural substrates i.e. fruits, potato, honey, molasses • Colour extracted is generally natural	• Prepared by dilution of synthetic acetic acid to 4%. • Carmel is added for colouring

Types of Brewed Vinegar

i. *Cider vinegar:* Made from apple juice. Should contain not less than 1.6% apple solids (with more than 50 % reducing sugars) and not less than 4 % acetic acid at 20° C.

ii. *Spirit vinegar:* Prepared from ethyl alcohol

iii. *Wine/ Grape Vinegar:* Prepared from Grapes/ wine

iv. *Malt Vinegar:* Malted barley and other cereals
v. *Sugarcane vinegar:* Sugarcane juice
vi. *Honey vinegar:* Prepared from honey
vii. *Potato vinegar:* Prepared form potato starch.

Methods/Steps

1. Alcoholic Fermentation

$$\underset{\text{Glucose}}{C_6H_{12}O_6} \xrightarrow{\textit{Saccharomyces cerevisiae}} \underset{\text{Ethyl Alcohol}}{2\,C_2H_5OH} + 2\,CO_2 + 55\text{ K Cal}$$

2. Acetous Fermentation

$$\underset{\text{Ethyl Alcohol}}{C_2H_5OH + O_2} \xrightarrow{\textit{Acetobacter aceti}} \underset{\text{Acetic Acid}}{CH_3COOH} + H_2O + 116\text{ K Cal}$$

Fruit
↓
Washing
↓
Grating
↓
Juice Extraction
↓
Alcoholic Fermentation
↓
Clarification
↓
Acetous Fermentation
↓
Raw Vinegar
↓
Ageing
↓
Clarification
↓
Packaging
↓
Pasteurization
↓
Cooling, Labeling and Storage

Flow Sheet for the Preparation of Vinegar

Grain strength : Unit for expressing acetic acid in vinegar.

Grain strength = acetic acid % in vinegar x 10 i.e. vinegar having 4.2% acetic acid has 42 grain strength.

Problems in Vinegar

Problem	Cause	Remedy
1. Wine flower	1. Formation of a yeast film on surface of unfermented juice on exposure to air. It causes cloudiness	1. Fill the carboys upto brim 2. Add 20-25% unpasteurized vinegar
2. Presence of Lactic acid bacteria	1. LAB interferes with acetification, cause cloudiness, produce lactic acid and mousy flavour	Use 20-25% unpasteurized mother vinegar or pure yeast culture

Points to Remember

√ Wine specifically refers to a fermented beverage prepared from grape juice unless otherwise specified generally containing 11-14% alcohol.

√ Still wine do not contain CO_2

√ Dry wines do not contain unfermented sugar

√ Fortified wines contain added alcohol eg vermouth

√ Brandy is an alcoholic distillate from fermented grape juice or wine unless otherwise specified.

√ Rum is distillate from alcoholically fermented sugarcane juice or molasses

√ Whiskey is distillate from fermented grains, rye, wheat.

√ Beer is prepared from fermentation of barley grains

√ Wine is pasteurized at 62.5° C for 20 min.

√ Vinegar is a condiment prepared by alcoholic fermentation of carbohydrates of the starch/ sugar containing commodities i.e. fruits, potato, molasses etc. or followed by acetous fermentation.

√ *Acetobacter aceti* is called as vinegar yeast

√ *Saccharomyces cerevisiae* is used in alcoholic fermentation

√ Grain strength = acetic acid % in vinegar x 10.

15

Enzymes, Waste Utilization and By Products

Enzymes are biocatalysts that change the rate of reaction without being changed themselves. All **(Most)** enzymes are proteins but all proteins are not enzymes. Ribonuclease is the enzyme which is non- protein in nature.

Properties of Enzymes

1. They increase the rate of a reaction but do not appear in the end products
2. They are very specific to pH
3. Cannot withstand high temperatures
4. They are substrate specific.

Importance : Enzymes speed up both the deteriorative as well as synthetic reactions.

Harmful Effects of Enzymes

- Enzymatic browning
- Phase separation in juices
- Softening of fruits

PPO Complex : Used as an indicator for blanching. It is a group of four enzymes viz., Polyphenol Oxidase, Phenol Oxidase, Polyphenolase, Phenolase.

Beneficial Effects of Enzymes

Enzyme	Function	Applications
amylase	Breakdown of starch	Manufacture of syrups, chocolates etc
Invertase	Inversion of sucrose into glucose and fructose	Manufacture of candies, confectioneries etc.
Protease	Breakdown of proteins	Baking, milling
Pectinase	Breakdown of pectin	Clarification of fruit juices, wines etc.
Glucose oxidase	Conversion of glucose in to gluconic acid	Desugaring of egg, potato etc.
Cellulase	Breakdown of cellulose	Clarification of fruit juices etc.
Glucose isomerase	Conversion of glucose into fructose	Manufacture of high fructose corn syrups
Lactase	Breakdown of lactose	Milk concentrate, ice cream, frozen desserts industry

Immobilized Enzymes

Enzymes are very useful in various food processing industries and are costly as well. Therefore, using the enzymes only once for catalyzing a reaction is not practically cost feasible and adds to the cost of production of a product. *Immobilization* is a technique by which enzyme is adsorbed / attached on to a non-reactive material and is added with the substrates during the reaction so that it is separable after the completion of the reaction and increase cost effeciency of use of enzymes.

Immobilized enzymes therefore are those that have been attached on a non-reactive material in such a way that they can catalyse the rate of reaction but are easily separable from the material (products) after completion of reaction and can be reused again and again for catalyzing same reaction.

Advantages of Immobilization

- Enzymes can be reused
- Enzymes become more stable during reaction
- Product is enzyme free
- Reduces cost of production
- Reaction kinetics may be enhanced.

Waste Utilization

Fruits and vegetables are highly perishable commodities and generate huge wastes (20-40%) often considered postharvest losses. Besides losses of fresh fruits and vegetables, wastes are also generated during processing of fruits and vegetables. These wastes may be in the form of peel, skin, seeds, pulper washing, kernel of nuts, cake, broken shell etc.

Problems with the Wastes

- Generate environmental pollution
- Become a source of contamination for other produce
- Requires additional costs for their disposal
- Can be responsible of many disease outbreaks
- Pollute water after rains

Why we should go for Waste Utilization?

- Reduce environmental pollution
- Generate additional income by producing useful byproducts
- Bulk of waste material is reduced
- Can be used for manufacture of useful products
- Disease outbreak can be avoided

Food Processing Waste Utilization and Manufacture of by Products

Crop	Waste (%)	Nature of Waste	By Products
Apple	20-30	Pomace	Juice, wine, vinegar, pectin, cattle feed etc.
Orange	50	Peel, Seeds, Pulp	Essential oils, pectin, cattle feed, peel candy etc
Lime	60	Peel, Seeds, Pulp	Essential oils, pectin, cattle feed, peel candy etc
Mango	40-60	Skin and Stone	Mango fat, bio colour
Mango Peels	12-15	Peel and pulp	Pectin, cattle feed, alcohol
Pulper waste	5-10	Fibre	Wine, vinegar, juice
Kernels	15-20	Hull and Kernel	Fat, tannins, starch
Pineapple	30-60	Peel, Core, trimmings, shreds	Juice, wine, syrup, cattle feed, biogas
Tomato	20-30	Core, peel, seeds	Animal feed, seed oil and meal
Potato	-	Peel, Coarse solids	Animal feed, single cell proteins
Banana	20-30	Peel	Animal feed
Wild apricot	60-70	Pulp, stone, kernel	Beverages, oil, cosmetics, processed products
Tamarind	-	Seeds	Starch for textile industry
Aonla	-	Fruit, pulp	Ayurvedic drugs
Pea	60-65	Pod	Animal feed
Cucumber	-	Seeds	Maghaz
Grape	-	Stem, pomace	Cream of tartar, jelly, chutney
Guava	-	Cores, seeds, peels	Guava cheese
Passion fruit	-	Rind, seeds	Pectin, oil
Pear	-	Peel, core	Animal feed, perry, vinegar

Points to Remember

- √ Enzymes are biocatalysts that the rate of reaction without being changes themselves.
- √ Ribonuclease is the enzyme which is non- protein in nature. Rest all enzymes are proteins in nature
- √ Enzymes are specific to pH and sunstrate
- √ Enzymes have both harmful as well as beneficial effects
- √ PPO Complex is used as an indicator for blanching
- √ Glucose oxidase converts glucose into gluconic acid while Glucose isomerase converts glucose into fructose
- √ Immobilization is a technique by which enzyme is adsorbed / attached on to a non reactive material and is added with the substrates during the reaction so that it is separable after the completion of the reaction.
- √ Immobilized enzymes are those that have been attached on a non reactive material in such a way that they can catalyse the rate of reaction but are easily separable from the material (products) after completion of reaction and can be reused again and again for catalyzing same reaction.
- √ Wastes generate environmental pollution and requires additional costs for their disposal
- √ Apple wastes can be utilized for preparation of fermented beverages
- √ Citrus peel is used for extraction of essential oil.

16

Spoilage of Fruits, Vegetables and Their Products

Spoilage may be defined as the deteriorative process which renders the food inedible or results into reduction of quality.

Spoilage is Mainly of Two Types

1. *Abiotic Spoilage:* Less potential for causing health hazards to consumers. Occurs due to
   - Action of enzymes (browning due to PPO)
   - Oxidation of fats (rancidity)
   - Putrification of proteins
   - Browning reactions between proteins and sugars
   - Physical changes (weight loss, shrivelling, colour fading etc.)
   - Caking, melting etc.
   - Absorption of moisture (hygroscopic foods)

2. *Biotic Spoilage:* More potential for causing health hazards. Associated with bacteria, yeasts, moulds etc. It may also be caused by insects, animals, rodents etc.

Two Types of Microbial Spoilage

a. *Spoilage by plant pathogens:* Appear on living tissues

b. *Spoilage by saprophytes:* Appear on dead and decaying tissues.

Factors Affecting Spoilage of Fruits and Vegetables

- *Moisture:* Higher the moisture more the spoilage. Microbial spoilage is reduced at moisture below 10%.
- *Temperature:* Higher the temperature (within limits) more the spoilage
- *pH:* Lower the pH lesser the spoilage and *vice versa*. Bacteria can not grow at low pH.
- *Composition of fruits and vegetables:* More the carbohydrates, higher are the chances of spoilage. Fats generally are not spoilt by microorganisms.

Spoilage Causing Microorganisms

1. *Bacteria:* Bacteria mostly grow between pH 4-8. The growth of bacteria at lower pH becomes difficult. Main groups of bacteria attacking fruits and vegetables

    a. Lactic acid bacteria: *Lactobacillus plantarum, L. brevis, L fermentum, Leuconostoc mesentroides, Pediococus sp.*

    b. Acetic acid bacteria: *Acetobacter aceti, Gluconobacter sp.*

    c. Coliform bacteria: *Escherichia coli, Enterobacter sp.*

    d. Spore forming bacteria: *Clostridium pasteurianum, Bacillus subtilis, Bacillus polymyxa*

    e. Pathogenic bacteria: *Salmonella sp, Shigella sp.*

2. *Yeasts:* These are mostly fermentative and produce carbon dioxide, alcohol, glycerol, acetaldehyde etc. Yeasts degrade starch, pectin, sugars etc.

   eg. *Candida* spp.
   Torulopsis spp.
   Rhodotorula spp.
   Saccharomyces spp.

3. *Moulds:* Various moulds are known to produce mycotoxins such as aflatoxins, patulins etc. Besides, they also produce compositional changes in the fruits and vegetables.

Important Fungi/Moulds found on Fruits and Vegetables

- *Alternaria*
- *Aspergillus*
- *Colletotrichum*
- *Rhizopus*
- *Sclerotia*

Mycotoxins : These are fungal metabolites highly toxic to many animals and potentially toxic to human beings.

Mycotoxin	Microorganism	Food products
Aflatoxins	*Aspergillus flavus*	Barley, corn
Patulin	*Penicillium sp.*	Apple, juices
Ochratoxin A	*Aspergillus* sp., *Penicillium* sp.	Corn, wheat, dough
Roquefortine	*Penicillium roqueforti*	Cheese

Aflatoxin – Aflatoxin is a naturally occurring **mycotoxin** produced mainly by two types of moulds i.e. *Aspergillus flavus* and *Aspergillus parasiticus*. *Aspergillus flavus* is found when certain grains are grown under stressful conditions such as drought. High temperature and high moisture favour growth of moulds. At least 13 different types of aflatoxins are produced in nature with aflatoxin B1 considered to be most toxic. Presence of *Aspergillus flavus* does not always indicate harmful levels of aflatoxin. It indicates only that the potential for aflatoxin production is present.

Classification of Microorganisms based on O_2 Requirement

Type	Molecular O_2 reqd.	Example
Obligate Aerobes	80-100	*Bacillus, Pseudomonas*
Facultative Anaerobes	50-100	*Salmonella, Streptococcus*
Facultative Aerobe	20-50	*Lactobacillus*
Obligate Anaerobe	0-20	*Clostridium*

Classification of Microorganisms based on Temperature Tolerance

Type	Temp. Requirements (°C)			Example
	Min.	Opt.	Max.	
Psychrophyllic	-5	8-10	20	*Pseudomonas, Micrococcus*
Mesophyllic	5	25-40	55	Yeasts, Molds, *Clostridium*
Thermophilic	20	50-55	80	*Bacillus stearothermophillus, Clostridium thermosaccharolyticum*

Classification of Microorganisms based on Tolerance to Salt and Sugar

1. *Halophilic* : Salt tolerant microorganisms

Bacteria	*Halobacterium*
Yeast	*Torulposis*
Moulds	*Aspergillus, Penicillium*

2. *Osmophillic* : Sugar loving microorganisms

Bacteria	*Bacillus*
Yeast	*Saccharomyces, Candida*
Moulds	*Aspergillus*

Food Spoilage

Food Illness : Any disease caused by consuming food.

Food Poisoning : It is the illness caused by consuming the poison present in the food. or illness caused by the poison present in the food when it is consumed. i.e. poison/toxin is produced first and then the food is consumed.

Food Infection : An illness caused by infection produced by invasion, growth, multiplication and damage to the host tissues by the pathogenic organisms carried by the food. Food is consumed first and then the poison/toxin is produced.

Difference between Food Poisoning and Food Infection

Food poisoning	Food Infection
• Poison / toxin is produced first and then the food is consumed	• Food is consumed first and then the poison / toxin is produced
• Caused by *Clostridium botulinum and Staphylococcus aureus*	• Caused by *Salmonella, Streptococcus, E. coil* etc.
• Toxin very potent and can put human life at risk	• Generally human life is not put at risk

Food Poisoning

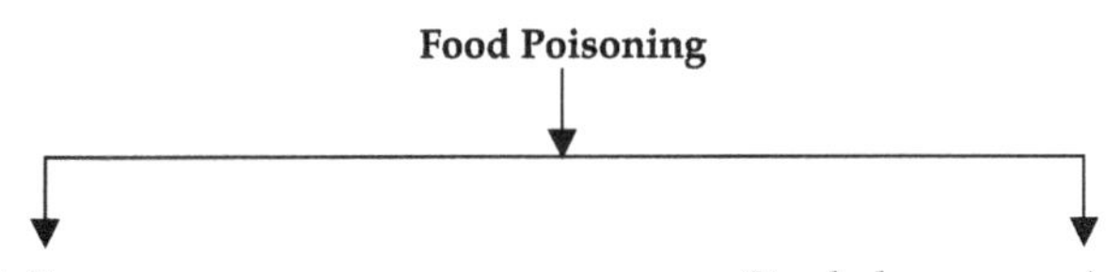

Botulism	Staphylococcus poisoning
1. Causal organism: *Clostridium botulinum*	1. Causal organism: *Staphylococcus aureus*
2. Toxin produced is very potent and a small quantity of it if consumed can cause mortality	2. Toxin not very potent
3. Destroyed by heating at 80° C for 30 min.	3. Destroyed at 121.1 ° C for 10 min.
4. Cause fatigue, digestive disturbances, dizziness, headache etc.	4. Cause salivation, vomiting, abdominal cramping etc.
5. Foods: Canned Foods, beans, Sweet corn, Asparagus	5. Meat, meat products, fish and fish products, milk etc.

Points to Remember

√ *Spoilage* may be defined as the deteriorative process which renders the food inedible or results into reduction of quality.

√ Oxidation of fats (rancidity) and putrification of proteins is abiotic spoilage

- √ Biotic Spoilage are more potential for causing health hazards and are associated with bacteria, yeasts, molds etc.
- √ Plant pathogens appear on living tissues and saprophytes appear on dead and decaying tissues.
- √ Bacteria mostly grow between pH 4-8. The growth of bacteria at lower pH becomes difficult.
- √ *Clostridium and Bacillus* are spore forming bacteria
- √ Mycotoxins are fungal metabolites highly toxic to many animals and potentially toxic to human beings.
- √ Aflatoxin are produced by *Aspergillus flavus* while *patulin* by *Penicillium sp.*
- √ Facultative anaerobes are basically aerobes but can grow occasionally under anaerobic conditions.
- √ *Bacillus stearothermophillus and Clostridium thermo-saccharolyticum* are thermophilic bacteria.
- √ Halophiles are salt tolerant and osmophiles are sugar loving
- √ In food poisoning poison/toxin is produced first and then the food is consumed while in food infection food is consumed first and then the poison/toxin is produced.
- √ Food poisoning is caused by *Clostridium botulinum and Staphylococcus aureus.*

17

Marketing and Export of Fresh and Processed Products

In a commercial horticultural enterprise, if the product is not marketed properly, safely and beneficially, all the investment, inputs, hard work goes in vain. Basically every commodity that is produced should reach the consumer and marketing is all about how the commodity reaches the consumer. There is no fun of producing any commodity if it does not reach the consumer.

The word *market* has been derived form Latin word "marcatus" meaning merchandise, where people conduct business. Market therefore refers to place where goods/ commodity and services are bought and sold.

Marketing is a sum total of business activities through which the goods and services are either bought or sold. Cundiff and Still defined marketing as "the business process by which products are matched with markets and through which transfer of ownership are affected.

Why Marketing of Fruits and Vegetables is Complex and Risky as Compared to Other Goods

- Fruits and vegetables are highly perishable in nature so require prompt distribution
- They are tender and are damaged and spoilt easily
- They require low temperature, high humidity (during handling, storage and distribution)
- They may also respond negatively to improper atmospheric conditions
- Packaging has to be given a great importance
- All the commodities can not be grown at all the places so the material is to be distributed to long distances.

Terminology used in Marketing (Swarup and Sikka, 1987)

Culled Fruits

Those fruits which may not have attained full colour, size and form typical to the variety, or bearing significant physical injuries (more than 3 pox marks due to hail) or healed insect damage, or bruised fruits even though they may be sound otherwise, are all known as culls. Drops, when the fruits have reached picking maturity, are also termed as culled fruits.

Marketable Surplus

It is the quantity of produce which is available for marketing.

Marketed Surplus

Marketed surplus is the actual quantity of the goods/ produce marketed.

Distributing Market

Market where the produce from the producing areas comes first and from where a large part of it is distributed to other markets eg. Azadpur Fruit and Vegetable Market, Delhi.

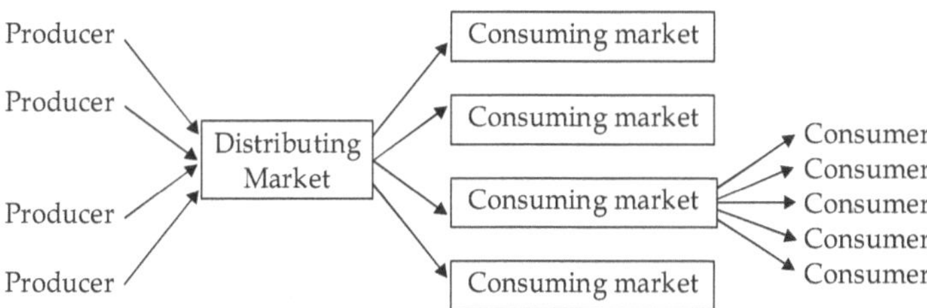

Consuming Market

A market which utilizes most of its supplies for local consumption is known as consuming market.

Assembling Point

Place where the growers assemble their fruit boxes, crates, baskets etc. for the purpose of transporting to various distributing and consuming markets.

Pre-harvest Contractor

Contractor who buys the standing crop before harvest from the growers and performs all the marketing operations from harvesting to final sale in a wholesale market at his own risk and cost.

Commission Agent

Commission agent is also known as 'Kacha Arhatia'. He performs the function of a seller (on behalf of the orchardists or pre-harvest contractor) for the produce booked to him by the growers. He charges commission for his services but does not take title to the goods.

Wholesaler

A wholesaler is a marketing mediator who buys and sells produce in bulk at his own risk. He takes title to the goods.

Wholesaler-cum-Commission Agent

A wholesaler-cum-commission agent is also known as 'Pucca Arhatiya' and performs the functions of commission agent as well as of wholesaler.

Retailer

A marketing functionary, usually licensed for sale of a particular category of products, who undertakes the job of retailing *i.e.,* to cater to the needs of consumers is known as retailer. He generally keeps a small establishment such as a shop with weighing equipments, secondary packaging material and facility for proper dislay of available goods.

Mashakhor

A sub-wholesaler who buys the produce either through the commission agent or from a wholesaler and sells to the retailers/consumers is known as mashakhor. He usually settles the price by negotiations.

Forwarding Agent

Forwarding agent arranges trucks and forward the produce to any desired destination, market and to the person for whom the produce has been marked by the consigner. Fee for the service is charged from the consignee, who in turn debits it to the account of the consigner.

Marketing Margin or Price Spread

Marketing margin or price spread refers to the difference between the price received (after deducting all marketing expenses incurred) by the grower and that paid finally by the consumer.

Marketing Channels between Growers and Consumers

Marketing channel may be defined as the route through which the produce/commodity moves from the producer to the consumer.

Basic Goals of Marketing Channel (Sharma 1991)

- The achievement of sufficiently broad product availability to assure ready exposure of the product to the potential purchasers
- The creation of an uninterrupted trading relationship with the reseller firms that comprise the system.

There May be a Number of Marketing Channels

a. Producer → Consumer

b. Producer → Retailer → Consumer

c. Producer → Whole saler → Retailer → Consumer

d. Producer → Agent → Retailer → Consumer

e. Producer → Agent → Whole saler → Retailer → Consumer

f. Producer → Forwarding agent → Commsssion agent → Whole saler
↓
Consumer ← Retailer

g. Producer → Producer's cooperative → Whole saler → Retailer
↓
Consumer

h. Producer → Pre harvest contractor → Commsssion agent →
Whole saler
↓
Consumer ← Retailer

i. Producer → Commission agent → Whole saler
(self as Forwarding agent)
↓
Consumer ← Retailer

j. Producer → HPMC/ APMC → Whole saler → Retailer
↓
Consumer

k. Producer → Processing unit → Commsssion agent → Whole saler
↓
Consumer ← Retailer

The more the number of middle man, the lesser would be the producer's share in consumer's rupee. In banana marketing for example the producers' share in consumer's rupee is around 31-35%.

Pre-harvest contractors at the time of 50 per cent crop maturity is opted by majority (88%) of the banana growers in Tamil Nadu. The major reason is that the pre-harvest contractors give advance payments to the farmers before harvest of the bananas to meet their immediate needs for production, consumption and social activities. Other reasons are poverty, high winds during monsoon causing damage to the plants, absence of institutional credits and crop insurance, small production and uneconomic quantities available for marketing, price discrimination lack of marketing knowhow and high marketing costs.

But, the pre-harvest contractors delay payments of balance amount after harvest and violate contracts if there is slump in prices.

Transportation of Fresh Fruits and Vegetables

Transportation is very important for moving the material from the producer's end to the market and then to the consumer. The following means are used for transportation of horticultural crops.

1. Non-refrigerated Trucks

It is the most common type of transportation system adopted for most of the horticultural crops and other goods in India. Fruit boxes are piled up in trucks which generally prefer to move during nights under less congested traffic on roads. This is a very prompt transportation system and generally used most commonly for transportation upto about 500 Km.

Advantages

- Trucks are independent and can move long distances
- Unlike railways they do not have to wait according to the timings of railway route. As soon as truck is filled it can move to its destination.
- Unlike rail and ships, they can reach different places within the cities.

Disadvantages

- More impact and bruising injury occurs to the fruit due to undulated roads and jerks experienced during the journey.
- Can create problems when truck faces some mechanical break down in a rural area where no repair facilities are available.
- More damage is caused during hot summers and rainy season to the fruits.

2. Rail

It is another very common type of transportation used for many of the crops and goods. Here also, the conditions are generally non-refrigerated and may affect the quality badly.

Advantages

- Less jerks and bruising to the fruit
- Suitable for long distances > 1000 Km
- Transportation can be faster than trucks.

Disadvantages

- Cannot reach at all places within a city
- Movement depends on clearance of route and schedule of trains.
- Trains do not reach most of the places in hills.

3. Jeep /Camper

This mean is becoming more popular when the destination market is very near i.e. 30-50 Km to the production centre.

Advantages

- Suitable for short distances.
- Can enter even the narrowest of roads, where trucks can not enter.

- Easy to own (less cost).
- Small growers can manage to hire a smaller vehicle rather than a larger vehicle such as trucks and rail.
- Transportation can be faster than trucks.

Disadvantages

- Not suitable for long distance transportation
- Costly for long distance transportation

4. Refer Vans

These are most suitable when the transportation is to be carried out at low temperatures commonly for the commodity which is very perishable i.e. flowers and leafy vegetables. Although, in India this mean of transportation is used over a limited extent, postharvest losses can be reduced to a considerable extend by making their use.

Advantages

- Better shelf life of produce
- Lesser postharvest losses

Disadvantages

- Costly and unaffordable for a common grower in India
- Not flexible in use. Cannot be used for all types of products as other traditional transportation systems. If used it would be very costly.

5. Ships

These are again used to vey less extent in India. This type of transportation is predominantly used during export of commodity in international markets.

Advantages

- Can carry very huge weights
- Lesser postharvest losses

Disadvantages

- Slow mean of transportation
- Not suitable for highly perishable produce
- Cannot reach everywhere unlike turcks and camper

6. Air Transport

Transportation through aeroplanes is common only for exports and for transportation to remote inaccessible areas within the country during natural calamities.

Advantages

- Better shelf life of produce
- Lesser postharvest losses
- Very fast means of transportation

Disadvantages

- Costly and unaffordable for a common grower in India
- Not flexible in use.
- Not available at all places
- Can not reach every small place

Palletization

This refers to packing the fruit/ vegetable boxes in huge containers / lots at airport or sea port before shipment or air transportation. Palletization facilitates shipment or air transportation due to clubbing of smaller units into a larger ones.

Most of the fruits and vegetables or even other goods transported by air are stored or transported as a pallet load. To minimize distribution costs, pallet loads should be established as soon as possible in the distribution chain and kept for as long as possible. ISO specifies two standard dimensions as the primary handling unit size for distribution of fresh fruits and vegetables within Europe for international trade.

a) 1200 x 1000 mm
b) 1200 x 800 mm

In sea transportation, bulk pack of potatoes are sent as containerized cargo in

a. Refrigerated containers (internal dimensions 5.364 x 2.255 x 2.255 m),
b. Ordinary general purpose containers with one door kept open for air circulation
c. Star vent containers with fitted fans for air circulation

IATA (International Aviation Transport Authority) pallet sizes are

a) IATA A 3070 mm x 2130 mm
b) IATA B 3070 mm x 2340 mm (*Source* Anon 2005)

Export of Fresh and Processed Fruits and Vegetables

There are several export promotion agencies working at national and state levels for increasing exports of various agricultural commodities (fresh as well as processed from India). Some of the important agencies in government and private sector are :

A. Government Sector

- Agricultural and Processed Food Products Export Development Authority (APEDA)
- Maharashtra State Agricultural Marketing Board, Pune
- Marine Products Export Development Authority
- Cashew Export Promotion Council of India
- Vegetable and Fruit Promotion Council, Keralam (VFPCK), Kerala
- Marketing Boards for tobacco, tea, coffee, and spices

B. Private Sector

- The Soybean Processors Association of India (soymeal),

- The Solvent Extractors Association of India (rapeseed meal, sunflowerseed meal, rice bran extraction),
- The Groundnut Extraction Export Development Association (peanut meal),
- The Indian Oil and Produce Exporters Association (peanut and sesame),
- The All India Cotton Seed Crushers Association (cottonseed meal),
- The All India Rice Exporters Association (rice),
- The Livestock and Meat Exporters Association (meat and meat products).

The Agricultural and Processed Food Products Export Development Authority (APEDA) was established by the Government of India under the Agricultural and Processed Food Products Export Development Authority Act passed by the Parliament in December, 1985. The Act (2 of 1986) came into effect from 13th February, 1986. The Authority replaced the Processed Food Export Promotion Council (PFEPC).

Functions

In accordance with the Agricultural and Processed Food Products Export Development Authority Act, 1985, (2 of 1986) the following functions have been assigned to the Authority. All the functions pertain to the scheduled products.

- Development of industries for export by way of providing financial assistance, undertaking surveys and feasibility studies, other reliefs and subsidy schemes.
- Registration of persons as exporters
- Fixing of standards and specifications for the purpose of exports
- Carrying out inspection of meat and meat products in slaughter houses, processing plants, storage premises, conveyances ensuring the quality of such products

- Improving packaging and marketing of products outside India
- Promotion of export oriented production & development of products
- Collection of statistics from the owners of factories or establishments engaged in the production, processing, packaging, marketing or export and publication of the statistics so collected or of any portions thereof or extracts therefrom
- Training in various aspects of the industries.

Products Monitored

APEDA has a mandate and responsibility of export promotion and development of the following scheduled products:

- Fruits, Vegetables and their Products.
- Meat and Meat Products.
- Poultry and Poultry Products.
- Dairy Products.
- Confectionery, Biscuits and Bakery Products.
- Honey, Jaggery and Sugar Products.
- Cocoa and its products, chocolates of all kinds.
- Alcoholic and Non-Alcoholic Beverages.
- Cereal, Cereal Products.
- Groundnuts, Peanuts and Walnuts.
- Pickles, Papads and Chutneys.
- Guar Gum.
- Floriculture and Floricultural Products
- Herbal and Medicinal Plants
- Rice (Non-Basmati)

In addition, APEDA also monitors exports of some non-scheduled items such as Basmati Rice, Wheat, Coarse Grains and also import of sugar.

(*Source:* www.apeda.com)

In order to meet out the national and international export demands of various agricultural products it is proposed to set up 60 Agri Export Zones in various states of the country.

Product		State	District/Area
Apples	1	Jammu & Kashmir	Districts of Srinagar, Baramula, Anantnag, Kupwara, Badgaum and Pulwama
	2	Himachal Pradesh	Shimla, Sirmour, Kullu, Mandi, Chamba and Kinnaur
Banana	3	Maharashtra	Jalgaon, Dhule, Nandurbar, Buldhana, Parbhani, Hindoli, Nanded and Wardha
Basmati Rice	4	Punjab	Districts of Gurdaspur, Amritsar, Kapurthala, Jalandhar, Hoshiarpur and Nawanshahar
	5	Uttarakhand	Udham Singh Nagar, Nainital, Dehradun and Haridwar
	6	Uttar Pradesh	Districts of Bareilly, Shahajahanpur, Pilibhit, Rampur, Badaun, Bijnor, Moradabad, J B Phulenagar, Saharanpur, Mujjafarnagar, Meerut, Bulandshahar, Ghaziabad
Cashewnut	7	Tamil Nadu	Cuddalore, Thanjavur, Pudukottai and Sivaganga
Chilli	8	Andhra Pradesh	Guntur
Coriander	9	Rajasthan	Kota, Bundi, Baran, Jhalawar & Chittoor
Cumin	10	Rajasthan	Nagaur, Barmer, Jalore, Pali and Jodhpur

contd...

contd...

Darjeeling Tea	11	West Bengal	Darjeeling
Dehydrated Onion	12	Gujarat	Districts of Bhavnagar, Surendranagar, Amreli, Rajkot, Junagadh and Jamnagar.
Flowers	13	Tamil Nadu	Dharmapuri
	14	Uttrakhand	Pantnagar Dehradun,
	15	Maharashtra	Pune, Nasik, Kolhapur and Sangli
	16	Karnataka	Bangalore (Urban), Bangalore (Rural), Kolar, Tumkur, Kodagu and Belgaum
	17	Tamilnadu	Nilgiri District
Flowers (Orchids) & Cherry Pepper	18	Sikkim	East Sikkim
Fresh & Processed Ginger	19	Assam	Kamrup, Nalbari, Barpeta, Darrang, Nagaon, Morigaon, Karbi Anglong and North Cachar districts
Gherkins	20	Karnataka	Tumkur, Bangalore Urban, Bangalore Rural, Hassan, Kolar, Chtradurga, Dharwad and Bagalkot
	21	Andhra Pradesh	Districts of Mahboobnagar, Rangareddy, Medak, Karinagar, Warangal, Ananthpur and Nalgonda
Ginger	22	Sikkim	North, East, South & West Sikkim
Ginger and Turmeric	23	Orissa	Kandhamal District
Grapes and Grape Wine	24	Maharashtra	Nasik, Sangli, Pune, Satara, Ahmednagar and Sholapur
Horticulture Products	25	Kerala	Districts of Thrissur, Ernakulam, Kottayaam, Alappuzha, Pathanumthitta, Kollam, Thiruvanthapuram, Idukki and Palakkod

contd...

contd...

Kesar Mango	26	Maharashtra	Districts of Aurangabad, Jalna, Beed, Latur, Ahmednagar and Nasik
Lentil and Grams	27	Madhya Pradesh	Shivpuri, Guna, Vidisha, Raisen, Narsinghpura, Chhindwara
Lychee	28	Uttranchal	Udhamsingh Nagar, Nainital and Dehradun
	29	West Bengal	Districts of Murshidabad, Malda, 24 Pargana (N) and 24 Pargana (S)
	30	Bihar	Muzaffarpur, Samastipur, Hajipur, Vaishali, East and West Champaran, Bhagalpur, Begulsarai, Khagaria, Sitamarhi, Sarannd Gopalganj
Mango & Grapes	31	Andhra Pradesh	Districts of Ranga Reddy, Medak & Parts of Mahaboobnagar Districts
Mango Pulp & Fresh Vegetables	32	Andhra Pradesh	Chittor District
Mangoes	33	Uttar Pradesh	Sahranpur, Muzaffarnagar, Bijnaur, Meerut, Baghpat, Bulandshar and Jyotifulenagar
	34	Maharashtra	Districts of Ratnagiri, Sindhudurg, Raigarh and Thane
	35	Andhra Pradesh	Krishna District
	36	West Bengal	Malda and Murshidabad
	37	Tamil Nadu	Districts of Madurai, Theni, Dindigul, Virudhunagar and Tirunelveli
Mangoes & Vegetables	38	Gujarat	Districts of Ahmedabad, Khaida, Anand, Vadodra, Surat, Navsari, Valsad, Bharuch and Narmada
	39	Uttar Pradesh	Lucknow, Unnao, Hardoi, Sitapur & Barabanki

contd...

contd...

Medicinal & Aromatic Plants	40	Uttranchal	Districts of Uttarkashi, Chamoli, Pithoragarh, Dehradun and Nainital
Medicinal Plant	41	Kerala	Wayanad, Mallapuram, Palakkad, Thrissur, Ernakulam, Idukki, Kollam, Pathanamittha, Thiruvananthapuram
Onion	42	Maharashtra	Districts of Nasik, Ahmednagar, Pune, Satara and Solapur
Oranges	43	Maharashtra	Nagpur and Amraoti
	44	Madhya Pradesh	Chhindwara, Hoshangabad, Betul
Pineapple	45	West Bengal	Darjeeling, Uttar Dinajpur, Cooch Behar and Jalpaiguri
	46	Tripura	Kumarghat, Manu, Melaghar, Matabari and Kakraban Blocks
Pomegranate	47	Maharashtra	Districts of Solapur, Sangli, Ahedabagar, Pune, Nasik, Osmanabad, Latur
Potatoes	48	Uttar Pradesh	Agra, Hathras, Farrukhabad, Kannoj, Meerut, Aligarh and Bagpat
	49	Punjab	Singhpura, Zirakpur (Patiala), Rampura Phul, Muktsar, Ludhiana, Jallandhar
	50	West Bengal	Districts of Hoogly, Burdwan, Midnapore (W), Uday Narayanpur and Howrah
Potatoes, Onion and Garlic	51	Madhya Pradesh	Malwa, Ujjain, Indore, Dewas, Dhar, Shajapur, Ratlam, Neemuch and Mandsaur
Rose Onion	52	Karnataka	Bangalore (Urban), Bangalore (Rural), Kolar
Seed Spices	53	Madhya Pradesh	Guna, Mandsaur, Ujjain, Rajgarh, Ratlam, Shajapur and Neemuch

contd...

contd...

Sesame Seeds	54	Gujarat	Amerali, Bhavnagar, Surendranagar, Rajkot, Jamnagar
Vanilla	55	Karnataka	Districts of Dakshin Kannada, Uttara Kannada, Udupi, Shimoga, Kodagu, Chickamagalur
Vegetables	56	Punjab	Fatehgarh Sahib, Patiala, Sangrur, Ropar & Ludhiana
	57	Jharkhand	Ranchi, Hazaribagh and Lohardaga
	58	West Bengal	Nadia, Murshidabad and North 24 Parganas
Walnut	59	Jammu & Kashmir	Kashmir Region – Baramulla, Anantnag, Pulwama, Budgam, Kupwara and Srinagar / Jammu Region - Doda, Poonch, Udhampur, Rajouri and Kathua
Wheat	60	Madhya Pradesh	Three distinct and contiguous zones:- Ujjain Zone comprising of Neemach, Ratlam, Mandsaur and Ujjain / Indore Zone comprising of Indore, Dhar, Shajapur and Dewas / Bhopal Division, comprising of Sehore, Vidisha, Raisen, Hoshangabad, Harda, Narsinghpur and Bhopal

POINTS TO REMEMBER

- √ The word *market* has been derived form Latin word "marcatus" meaning merchandise
- √ Market is place where goods/commodity and services are bought and sold.
- √ *Marketing* is a sum total of business activities through which the goods and services are either bought or sold.

- √ *Assembling Point* is defined as the place where the growers assemble their fruit boxes for the purpose of transporting to various distributing and consuming markets.
- √ *Pre-harvest Contractor* is one who buys the standing crop before harvest from the growers and undertakes to perform all the marketing operations from picking to final sale in a wholesale market at his own risk and cost.
- √ *Wholesaler* is one who buys and sells produce in bulk at his own risk. He takes title to the goods.
- √ Mashakhor acts as a sub-wholesaler
- √ *Marketing channel* may be defined as the route through which the produce / commodity moves from the producer to the consumer.
- √ The more the number of middle man, the lesser would be the producer's share in consumer's rupee.
- √ In road transport there is more impact and bruising injury to the fruit while in rail transport impact injury is very less.
- √ Jeep / Camper is suitable for transportation to nearby markets 30-50 Km.
- √ Refer vans are most suitable for very perishable commodities i.e. flowers and leafy vegetables.
- √ Palletization refers to packing the fruit/ vegetable boxes in huge containers / lots at airport or sea port before shipment or air transportation.
- √ APEDA stands for Agricultural and Processed Food Products Export Development Authority
- √ APEDA, MPEDA and Tea board are govt. agencies for export promotion.
- √ In order to meet out the national and international export demands of various agricultural products it is proposed to set up 60 Agri Export Zones in various states of the India.

18

Planning of Fruit and Vegetable Processing Industry

Before beginning the actual work on the establishment of a Fruit and Vegetable Processing Industry the following points should be considered:

- *Availability and possibilities of obtaining Funds:* Investment of own money or obtaining loans from banks etc.
- *The scale of industry (Large, Small, Cottage or Home):* Has to be decided based on availability of raw material, funds availability, land, labour and other resources.
- *The location:* The processing unit should be located near to the actual production site of raw material so that the transportation costs of high bulk raw materials is reduced. There is no fun or justification of establishing apple processing industry in Lucknow or Bangalore and mango processing industry in Shimla as the related raw materials would have to be transported through long distances.

- *Site:* Site selected in the location should be sunny area free from dampness and sufficient space for waste disposal should be ensured.
- *Availability of raw materials:* Raw material should be available near to the processing unit throughout the year in order to have maximum capacity utilized.
- *Modalities of operation:* How the plant would be made to run throughout the year? What would be the sequence of products which would be prepared during the on and off season of the raw materials?
- *Types of products to be prepared:* Whether the unit would prepare single product or many products? What would be the share of each of the products?
- *Possible markets for the manufactured products:* What would be the destination market of the products manufactured?
- *Availability of labour/ inputs etc.:* How the labour would be employed? How many labourers would be employed? How many would be retained during the off season?
- *Availability of water:* Availability of 24 hr potable water supply should be ensured.
- *Scope for expansion:* In future if the business runs successfully whether there exists any scope for the expansion of plant?
- *Transportation facilities:* What would be the mode of transportation for raw materials, inputs and finished products?
- *Distance from markets:* What cost would be incurred on transportation of the material?
- *Environment/Pollution etc.:* What measures would be taken to reduce environmental pollution generated due to processing wastes?

Points to be Considered while Selecting a Factory Site

- Availability of raw material in nearby areas.
- Transportation facility
- Environment
- Scope for expansion

Factory Building

- The building may be single storeyed or multi storeyed.
- Flooring should be firm
- Heavy machinery should be grouted at ground floor
- A slope of $1/4^{th}$ of a inch per foot is necessary in the processing hall.
- All rooms and halls should be well lighted and ventilated.
- All windows and ventilators should be provided with fine wire gauge to prevent entrance of flies, wasps and insects.
- Roof should be high more then 10 feet for home scale industry and > 14 feet preferably and well ventilated to provide outlet (exhaust fan) for vapours and steam.
- A part of roof may be made of glass so as to permit entrance of light.
- Sufficient number of dressing and toilet rooms/wash basins should be provided and for males and females separately.
- The interiors of the processing area should be painted with washable paint minimum upto a height of 5 feet.

Water Supply

- Sufficient supply of potable water.
- Water should not be alkaline or very hard and should be free from organic matter.

- Presence of iron and sulphur in water makes it unfit for preparation of brines, syrups and other processed products.
- Saltish water would also affect the taste of the products so it should also be avoided.
- Water quality should be got tested from a reputed laboratory for obtaining FPO licence.

Labour/Staff

- All Labour and staff should wear clean cloths, aprons, caps, scarf and gloves.
- They should be examined at regular intervals and should be vaccinated against infectious and contagious diseases.
- The chemist and supervisors should be well qualified and trained to ensure the desired standard of production and quality.

Points to Remember

√ Processing unit should be located near to the place where raw material is produced.

√ 24 hr potable water supply should be ensured

√ Labour and staff should wear clean cloths, aprons, caps, scarf and gloves and be examined at regular intervals and should be vaccinated against infectious and contagious diseases.

√ A slope of 1/4th of a inch per foot is necessary in the processing hall.

√ All rooms and halls should be well lighted and ventilated.

√ All windows and ventilators should be provided with fine wire gauge to prevent entrance of flies, wasps and insects.

√ Sufficient number of dressing and toilet rooms/wash basins should be provided and for males and females separately.

- √ The processing unit should be connected to road for the speedy and low cost transportation of raw material to the factory and finished products from the factory.
- √ FPO licence must be obtained form Ministry of Food Processing Industries, Govt., of India before starting actual production of the products.
- √ Products should be always be prepared under the guidance and supervision of a trained staff.
- √ All the prepared products should be truthfully labeled.

PART III

Novel Technologies

19

Novel Technologies in Food Preservation

With the increasing health consciousness of consumers, there has been an increasing concern over the use of thermal processing methods like blanching, pasteurization, sterilization etc. for enhancing the shelf life of foods as these methods fail to preserve the fresh like nutritional, therapeutical and sensory qualities of the food. With the increased consumer demand for fresh and fresh-like foods, interest in minimal processing of foods is increasing. The trend is also shifting from the traditional heat processing to those technologies that employ no or minimum heat application, thereby retaining the quality to a maximum level. One of the ways to accomplish this is via use of non-thermal processes. Over the past few years a number of non-thermal processes or cold preservation methods have been invented and developed. These methods are being evaluated in pasteurization and sterilization of various foods or in the extension of shelf life of various packaged foods. Some of such novel technologies and the products prepared, have been discussed here under.

1. Extrusion

Extrusion is a size enlargement process wherein small granular food or powdered particles are re-formed into larger pieces with different shapes, textures, colours and flavours from the basic ingredients. In *extrusion* food is compressed and forced to pass through a fine opening (die). As soon as it comes out of the opening, the outside pressure is very less as compared to the pressure inside the screw of extruder and the food expands instantaneously. If the food is heated the process is called as *extrusion cooking*. It is a process which combines several unit operations including cooking, mixing, kneading, shearing, compressing, shaping and forming. Extrusion adds a variety to the products by changing shape, size, texture and even some of the minor ingredients of the food and at a reduced cost than most of the other processing methods. Extrusion cooking (not extrusion) is a high temperature short time (HTST) process that reduces microbial contamination and inactivates enzymes.

Theory

During extrusion cooking of starch based foods (maize, wheat etc.), the moisture content is increased by added water. The starch molecules absorb water swell and become gelatinized. The macromolecular structure of starch molecules opens up and a viscous plasticized mass is produced (Mercier 1980). While, during extrusion cooking of protein-based foods (e.g. soy meal and defatted oil seed flours) the quaternary structure of the protein opens in the hot moist conditions to produce a viscous plasticized mass.

The main factors influencing the nature and quality of extruded product are the operating conditions (such as temperature, pressure, diameter of die aperture, shear rate etc.) and the rheological properties of food (moisture content, physical state and chemical composition like fats, proteins, starch, sugars, etc.).

Principles of preservation : In both the processes i.e. cold extrusion and hot extrusion, the main mode of preservation is the low water activity.

Equipment

The equipment called extruders may be classified on the basis of

a. Construction i.e. single-screw or twin screw extruders
b. Operation i.e. cold extruders or hot extruders/extrusion cookers.

The working of extruders is basically the same in all types. The feed material is fed into the extruder barrel. The material is conveyed by screw(s), compressed and worked (shearing and kneading) to transform the granular feed material into semi solid plasticized mass. The food is forced through a restricted opening (the die) at the discharge end of the screw. As soon as the food is released out through the die aperture, the material comes to a low pressure area (out side the screw barrel) from a region of high pressure (inside the screw barrel), causes instantaneous expansion of steam and gas in the material which reduces the density of the product formed. The expanded food is cut by rotating knives into variety of shapes like rods, spheres, doughnuts, tubes, strips, squirls or shells. The moisture is then removed by evaporation from expanded product (Fellows, 1988).

Types of Extruders

Various types of extruders as explained by Fellows are

1. *Hot extruder/Extrusion cooker*

Food is heated by

(a) Steam circulating through a jacketed barrel, or
(b) Electric induction heating elements which directly heat the barrel and
(c) Heat generated due to friction between screw ribs and the barrel walls and food.

The compression of food is achieved by the decreasing screw pitch in a tapered barrel, often supplemented by reducing the aperture by alteration of die position. The extent of expansion depends upon the pressure and temperature generated in the extruder and the rheological properties of food. A low pressure or large die aperture are used to produce high density products like preforms or half products which can be converted into final product by frying, toasting or puffing.

2. *Cold Extruder*

In this type, the feed is extruded into strips without cooking or distortion of food by expansion. The screw operates at low pressure with little friction. Used to produce pasta, hot dogs, pastry doughs, confectionary etc.

3. *Co-extruders*

Here the extruder is fitted with a special die to inject a filling into an outer shell continuously. It may be used to produce products like filled confectionery and may be combined with both the hot or cold extruders.

4. *Single Screw Extruders*

This extruder has a feed section which compresses feed to a homogenous mass, a kneading section to compress, mix and shear the plasticized food and a cooking section. There are 3 types of flows (a) drag flow (material flows forward due to dragging of feed by screw flights), (b) pressure flow (by building up of pressure due to material movement) and (c) leakage flow (backward movement along the barrel).

Advantages : Low capital cost, low operating cost, require less skill to operate.

Disadvantages : Not suitable for handling, oily, sticky or very wet material, limited to a specific range of granular particle sized feed.

5. *Twin Screw Extruders*

There are two screws which rotate within a '8 shaped' bore in the barrel. The material is intermeshed and mixed properly due to movement of 2 screws and therefore improves the efficiency.

Advantages : Prevents rotation of material inside barrel, suitable for handling oily, sticky and wet material, suitable for a range of feed types from fine powders to grains

Disadvantages : High capital and operational costs

Operating Conditions

These may vary in different types of extruders and depending upon the feed material used. General conditions are as follows for different equipments (Harper, 1979).

Field moisture	11 - 28%
Product moisture	2 - 25%
Maximum product temperatures	52-199°C
Pressures	1500-70,000 KPa
Screw speeds	30-450 rpm
Residence time	15-90 sec

Applications

Extrusion may be used for preparation of different products like, breakfast cereals, crispbread, cornflakes, dough pellets, texturized soya products, confectionery products like fruit gums, wine gums, snacks, pasta products, pet foods, animal feeds, sausage products, etc.

With the increase in urbanization and the preference of consumers to the ready to eat and ready to cook, type of products, the extruded products are getting popularized very fast due to the merits of the technology. However, some of the problems which are often encountered during the process are

Limitations

- Fading of product colour due to expansion on excessive heat.
- Loss of flavour due to their volatile nature.
- Maillard browning and loss of proteins
- Loss of vitamins (upto about 50%)

The use of HTST helps in better retention of nutritional and sensory quality attributes.

Points to Remember

- √ Extrusion is a size enlargement process
- √ If the food is heated the process is called as *extrusion cooking*.
- √ Extrusion cooking (not extrusion) is a high temperature short time (HTST) process that reduces microbial contamination and inactivates enzymes.
- √ The main mode of preservation in extrusion is the low water activity.
- √ There are 3 types of flows (a) drag flow (material flows forward due to dragging of feed by screw flights), (b) pressure flow (by building up of pressure due to material movement) and (c) leakage flow (backward movement along the barrel).
- √ Pressures of 1500-70,000 KPa and temperatures of 52-199° C are generated during extrusion.
- √ Extrusion may be used for preparation of breakfast cereals, crispbread, cornflakes, dough pellets, texturized soya products, confectionery products like fruit gums, snacks, pasta products, pet foods, animal feeds, sausage products, etc.

2. Irradiation

Before discussing irradiation process it is a pre-requisite to understand some of the terms, definitions and units associated with it. Some important of these are discussed hereunder :

Becqeral (Bq) : One unit of disintegration per second

Curie (Ci) = 3.7 x 10^{10} Bq

Half life : Time taken for the radioactivity of a sample to fall to half its initial value

Electron volt (eV) : Energy of radiation (usually mega electron volts (MeV), 1 eV = 1.602 x 10^{-19}J). One ev is the energy gained by an electron in moving through a potential difference of 1 volt.

Grays (Gy) : Absorbed dose (where 1 Gy = 1J of energy absorbed per kilogram of food). Previously rads was used where 1 rad = 10^{-2}J kg^{-1}. 1Gy = 100 rads.

Radiation Source : Only γ-rays from Cobalt-60 or Caesium–137, x-rays generated by a machine at a maximum energy of 5 mega volts; or electrons generated by a machine at a maximum energy of 10 mega volts can be used which are too low to induce radio activity in any material including food, exposed to them.

Radiation Dose : The quantity of radiation energy absorbed by a food, during treatment. It is measured in Gy or rads.

Election Beam : Streams of electrons (β particles).

Ionizing Radiation : High energy radiations that can penetrate other atoms and produce electrically charged particles called ions. It includes γ-rays, x-rays, cathode rays, microwaves, radiowaves, ultrasonics, laser and electronic beams. Non-ionizing radiation include infra red rays, UV rays and hertzian rays.

Radioactivity : Property of spontaneous release of radiation energy from an atom having unstable nucleus.

Radappertization : It is equivalent to *radiation sterilization* / commercial sterility. It is the destruction of practically all microorganisms achieved by high doses of irradiation.

Radicidation : It is equivalent to pasteurization of milk. It is a low dose irradiation treatment to destroy viable non-spore forming pathogenic microorganisms, to reduce or eliminate particularly *Salmonella*. Spores of *Clostridium botulinum* and *C. perfringens* are not destroyed with radicidation.

Radurization : Low doses of radiation to destroy a sufficient number (most) of microorganisms and enhance storage life of different types of foods is called radurization or radiation pasteurization.

Thermoradiation : This is a simultaneous or sequential combination treatment of heat and radiation, useful for food preservation. Heat and irradiation have synergistic effects when used together for killing spores of anaerobes e.g. *Bacillus stearothermophilus*, D value of 210 K rads at 27°C is reduced to 157 K rads when temperature is increased to 60°C.

Picowaved : This term is used to label the food which has been treated by low levels of ionizing radiations.

Food Irradiation

The deliberate treatment / exposure of food to ionizing radiations such as γ-rays, x-rays or electron beam, from a radioactive or machine generated source, under controlled conditions, to disinfest, sterilize and preserve the food or prevent underiable changes i.e. sprouting and prolong its shelf life is known as *Food Irradiation*. Foods can be irradiated wet, dry, thawed or frozen. Since the process does not involve generation of heat, it is also called *Cold Sterilization*. Food can be pasteurized or sterilized without causing changes in freshness and texture and the efficiency of the process makes it suitable to treat even the pre-packed foods.

Theory

The process of irradiation involves passing of food through a radiation field under controlled conditions for a specific period of time and, allowing it to absorb desired radiant energy. The food itself never comes in direct contact with the radioactive material, so, the irradiated food is not in any way made radioactive.

The ionizing radiations i.e. γ-rays and electrons when absorbed by food break the chemical bonds. The process of ionization forms electrically charged ions or neutral free radicals, which further react to cause changes in an irradiated material known as *Radiolysis*. These reactions cause destruction of micro-organisms, insects and parasites during irradiation. Water present in the food is ionized by radiation. Electrons are expelled from water molecules and break chemical bonds. The products then recombine to form hydrogen, hydrogen peroxide, hydrogen radicals (H^o), hydroxyl radicals (OH^o) and hydroperoxyl radicals ($H_2O_2^o$). The process occurs in two steps (Hughes 1982, Robinson 1986).

1. Ionization of water : $H_2O \rightarrow H_2O^+ + e^-$
   $e^- + H_2O \rightarrow H_2O^-$
   $H_2O^+ \rightarrow H^+ + OH^o$
   $H_2O^- \rightarrow H^o + OH^-$
2. Formation of free radicals : $H^o + H^o \rightarrow H_2$
   or $OH^o + OH^o \rightarrow H_2O_2$
   or $H^o + OH^o \rightarrow H_2O$
   or $H^o + H_2O \rightarrow H_2 + OH^o$
   or $OH^o + H_2O_2 \rightarrow H_2O + HO_2^o$
   $H^o \rightarrow + O_2 HO_2^o$

These free radicals are extremely short lived (10^{-5} sec) and are highly reactive, therefore combine readily with other molecules to form stable products or again free radicals. The radicals are sufficient to kill the bacterial cells. Fat soluble components and essential fatty acids are lost during irradiation and such products (e.g. dairy products) are therefore unsuitable to irradiation due to development of rancid off-flavours which is also accelerated by the presence of oxygen, therefore some of the meat products are irradiated under vacuum.

Equipment

An irradiation chamber, shielded by 1.5 to 1.8 m thick concrete wall, is used to carry out irradiation. The radiation comes from a high-energy isotope source producing γ-rays from Cobalt-60 or Caesium-137 or less commonly, a machine source producing high energy electron beams. The Cobalt-60 emits γ-rays at two wavelengths having 1.17 MeV and 1.33 MeV energies. The activity of Cobalt-60 and Caesium-137 is rated at (222-370) x 10^{10} Bq g^{-1}. An isotope source cannot be switched off, therefore, is shielded with a pool of water below the process area, and during operation, the source is raised and packed food, by conveyers is transported through the radiation field in a circular path or is turned around their own axis, so that the contents are irradiated uniformly through all sides. Cobalt-60 has a half life of 5.26 years and therefore, requires replacement of 12.3% of the activity each year to retain the desired output of the plant. For economic operation of the plant continuous processing with a continuously decaying source, is desirable.

On the other hand machine sources which are basically electron accelerators, consist of a heated cathode to supply electrons which are accelerated by a high voltage electrostatic field. Either the electrons or the produced x-rays are used directly on food. The advantages of machine source are

- They can be switched off as desired after completion of treatment
- Electron beams can be diverted over the packaged food to ensure even-dose distribution
- Handling is simple and involves less risk than that of radioactive source.

The disadvantage however is that the equipment is expensive and relatively inefficient in producing radiation.

Mode of Action on Micro-organisms

The main mechanisms by which the micro-organisms are killed are

- Effect on DNA and RNA molecules of microorganisms. The DNA double helix fails to unwind and micro-organisms cannot reproduce
- Change the structure of cell membranes
- Affect metabolic enzyme activity.

However, the spore forming bacteria and those which can readily repair the damaged DNA are more resistant to irradiation than vegetative cells and non-spore formers. Viruses are also resistant and are generally not affected by the dose levels used in commercial food processing.

Applications

The maximum recommended dose for treating foods is 15-70 KGy with the average dose not exceeding 10K Gy. The important applications of irradiations are

1. Low Dose (< 1K Gy)

- Disinfestation of grains and fruits for insects and larvae (<1 KGy)
- Sprout inhibition in bulbs and tubers i.e. potatoes, onion and garlic (0.03-0.20 K Gy)
- Delay of fruit ripening (0.25-0.75 K Gy)

2. Medium Dose (1- 10 KGy)

- Sterilization of herbs and spices (8-10 KGy)
- Reduction of pathogens i.e. *Salmonella, Campylobacter, Escherichia* or *Vibrio* sp (2.5-10 KGy)

- Extension of shelf life of fruits, vegetables and mushrooms, reduction of spoilage microorganisms in meat poultry and sea foods (1.5 - 3 KGy).

3. High Dose (> 10 KGy)
   - Sterilization of packaging material 10-25 KGy
   - Sterilization of packaged meat, poultry products to make them shelf stable without refrigeration (25-70 K Gy)
   - Sterilization of hospital diets (25-70 K Gy)

At commercial dose levels, irradiation does not cause greater damage to nutritional and sensory quality as compared to that caused by other commercial preservation and processing methods. However, at higher dose levels, chemical changes in the structure of amino acids and proteins causes changes in aroma and taste of foods. The order of sensitivity of various vitamins in decreasing order is thiamin > ascorbic acid > pyridoxin > riboflavin > folic acid > cobalamin > nicotinic acid. Vitamins D and K are practically unaffected while vitamins A and E undergo some losses.

Detection of Irradiation in Foods

Since irradiation produces no major physical, chemical or sensory changes in the food when treated at or below the commercially recommended doses, the detection of irradiation becomes important and focuses upon minor changes in chemical composition, physical or biological changes in the food. Some important methods for detection of irradiation have been given as under :

1. *ESR (Electron Spin Resonance Spectroscopy)* : This detects the radicals produced by irradiation, which are stable in solid or dry food components such as bones in meats and fish products.
2. *TL (Thermo Luminescence)* : Energy is trapped in the crystalline lattice (containing minerals in fruits and

vegetables) during irradiation. When food is heated in controlled way this stored energy is released as light and measured by a sensitive photon counter.

3. *PSL (Photo Stimulated Luminescence)* : This method is similar to the TL but here pulsed infra-red light is used to release the stored energy in food instead of heat. Also, the isolation of minerals is not needed as in the case of TL, therefore, is more efficient and less laborious.

4. *Detection of 2-alkyl cyclobutanones (2-CBs)* : The detection of 2-CBs (the radiolytic products formed from fatty acids), by mass spectrometry is used to detect irradiation in meat products and fruits i.e. mango, papaya and avocado. TLC and HPLC can also be used to detect 2-CBs.

5. *Detection of DNA fragmentation* : If the food has not been at any time heat treated, the detection of irradiation can be done by recording DNA fragmentation by microgel electrophoresis or ELISA (Enzyme linked immunosorbant assay).

6. *Biological methods* : The measurement of both dead and live microorganisms in irradiated food by Limulus Amebocyte Lysate in conjunction with Gram negative bacteria (LAL/GNB) and Direct Epifluorescent Filter Technique with an aerobic plate count (DEFT/APC) gives the information about the number of microorganisms destroyed by irradiation. If DEFT count exceeds the APC by 10^4 or more, the food is considered to be irradiated.

7. *Other methods* : Measurement of physical changes by electrical impedance, changes in viscosity, electric potential, nuclear magnetic resonance, near-infra red spectroscopy, capillary gel electrophoresis to detect the changes in the size of proteins and peptides are some of the new methods which are also being tried to detect irradiation in various food products.

Points to Remember

- √ 1 Gy = 1J of energy absorbed per kilogram of food
- √ 1 rad = 10^{-2}J kg^{-1}. 1Gy = 100 rads.
- √ Only γ-rays from Cobalt-60 or Caesium-137, x-rays generated by a machine at a maximum energy of 5 mega volts; or electrons generated by a machine at a maximum energy of 10 mega volts can be used which are too low to induce radio activity in any material including food, exposed to them.
- √ *Radappertization* : It is equivalent to *radiation sterilization* / commercial sterility. It is the destruction of practically all microorganisms achieved by high doses of irradiation.
- √ *Radicidation* : It is a low dose irradiation treatment to destroy viable non-spore forming pathogenic microorganisms to reduce or eliminate particularly *Salmonella*. Spores of *C. botulinum* and *C. perfringens* are not destroyed with radicidation.
- √ *Radurization* : The use of low doses of radiation to destroy a sufficient number of microorganisms and enhance storage life of foods.
- √ *Thermoradiation* : This is a simultaneous or sequential combination treatment of heat and radiation, useful for food preservation.
- √ *Picowaved* : This term is used to label the food which has been treated by low levels of ionizing radiations.
- √ Food Irradiation is the deliberate treatment / exposure of food to ionizing radiations such as γ-rays, x-rays or electron beam, from a radioactive or machine generated source, under controlled conditions, to disinfest, sterilize and preserve the food prevent undesirable changes i.e. sprouting or prolong its shelf life
- √ Food Irradiation is also called *Cold Sterilization.*
- √ Free radicals are extremely short lived (10^{-5} sec) and are highly reactive, therefore combine readily with other molecules to form stable products or again free radicals. The radicals are sufficient to kill the bacterial cells.
- √ Fat soluble components and essential fatty acids are lost during irradiation and such products (e.g. dairy products) are therefore

unsuitable to irradiation due to development of rancid off-flavours which is also accelerated by the presence of oxygen, therefore some of the meat products are irradiated under vacuum.

√ The main mechanisms by which the micro-organisms are killed are (i) effect on DNA and RNA molecules of microorganisms. The DNA double helix fails to unwind and microorganisms cannot reproduce, (ii) Change the structure of cell membranes, (iii) Affect metabolic enzyme activity.

√ *Low dose* (< 1K Gy) are used for disinfestation of grains and fruits for insects and larvae.

√ High dose (> 10 KGy) are employed for sterilization of packaging material and packaged meat, poultry products to make them shelf stable without refrigeration (25-70 K Gy) and sterilization of hospital diets (25-70 K Gy)

3. Ohmic Heating

The ohmic heater uses resistance heating within the flow of electrically conducting liquid and particulates. Temperature of liquid and well as particulates increases rapidly to kill the microorganisms and we can handle food products containing particulates of upto 25 mm by this system.

Theory

Ohmic heating takes place when an electric current, usually a low frequency alternating current (50-60 Hz), is passed through a conducting medium i.e. brine. In similarity to microwave heating, the electrical energy is transformed into thermal energy during ohmic heating. However, unlike microwave heating the depth of penetration is virtually unlimited. The generation of heat is governed by spatial uniformity of electrical conductivity throughout the product and its residence time in the ohmic heater. A large temperature gradient does not exist within the food because, liquid and particulates are heated virually simultaneously.

Equipment

The ohmic heater is a column consisting of four or more electrode housings machined from a solid block of polytetra fluoro ethylene (PTFE) and encased in stainless steel, each containing a single cantilever electrode. The column is mounted in vertical position with flow of product in an upward direction. The electrode housings are connected using stainless steel interconnecting tubes which are lined with an electrically insulating plastic liner i.e poly vinylidene fluoride (PVDF) and glass. Each heating section has the same electrical impedance (Crepaco, 1992).

Applications

The applications of ohmic heating are dependent upon the electric conductivity of the product or liquid media. Since most food preparations have moderate quantity of free water, dissolved ionic salts and hence conduct sufficiently well, the process is applicable for a wide variety of food products. However, the system does not directly heat fats, oils, alcohols, bone or crystalline structures such as ice.

Advantages

1. Less heat damage to the liquid phase and prevents surface overheating/overcooking of particulates.
2. No heat transfer surfaces that reduce the possibility of deposit formation and transfer of burnt particles to finished product.
3. There is no need for mechanical agitation so particle structure is not damaged.
4. Causes high degree of damage to microorganisms and provides better retention of nutritional quality due to almost instantanous heating.
5. Process can be combined easily with aseptic packaging and processing. Such products can be stored at ambient temperatures for long durations.

POINTS TO REMEMBER

- √ Ohmic heater uses resistance heating within the flow of electrically conducting liquid & particulates of size upto 25 mm.
- √ Usually a low frequency alternating current (50-60 Hz) is used in ohmic heating
- √ In similarity with microwave heating, the electrical energy is transformed into thermal energy during ohmic heating.
- √ Unlike microwave heating the depth of penetration is virtually unlimited. The generation of heat is governed by spatial uniformity of electrical conductivity throughout the product and its residence time in the ohmic heater. Product does not experience a large temperature gradient within itself as, liquid and particulates are heated virtually simultaneously.
- √ Ohmic heating does not directly heat fats, oils, alcohols, bone or crystalline structures such as ice.
- √ There is less heat damage to the liquid phase and prevents surface overheating/overcooking of the particulates.

4. Microwave Processing and Dielectric Heating

With the increasing consumer demand for foods that offer freshness, greater convenience and time saving, microwave processing has come up as a strong substitute to conventional thermal processing in recent years. The use of microwaves, offer rapid and economic methods for preparation of different types of food products having high organoleptic and nutritional value.

Energy Type and Wavelength*

Cosmic rays	Very short
Alpha, beta, gamma rays	< 100 nm
X rays	100-150 nm
UV	13.6-400 nm'
Visible	400-800 nm
Infra red	800 nm -10 mm
Micro waves	30 cm
Dielectric	25 m
Radio waves	3 0m - 3 Km

* These wavelengths may not be very exact but would give a comparative idea of different types

Theory

Microwaves are a form of electromagnetic energy produced at specific frequency bands (2450 MHz or sometimes 896 MHz in Europe and 915 MHz in USA), generated by magnetron (a device that converts electric energy at low frequencies into an electromagnetic field), with centers of positive and negative charge that change direction billions of times per second. This causes the instantaneous penetration and heating of foods. The molecular structure of water consists of a negatively charged oxygen atom and positively charged hydrogen atoms. This forms an electric dipole. When rapidly oscillating electric field is applied to food, dipoles reorient with each change in the field direction. Heat generation is caused by (i) rotation of dipoles (Dipole polarization) and (ii) oscillation of ions (electron polarization) in the food with the alternating field. In microwaves, a magnetic field and an electric field orient perpendicularly to each other. The electric field promotes rotation of polar molecules, resulting in heat generation by molecular friction. The various distortions and deformations to the molecular structure, caused by realignment of the dipoles, dissipate the applied energy as heat.

When food is placed in the path of microwaves electromagnetic energy is absorbed and converted to heat. Food is heated by generation and deposition of heat in the food itself, unlike that in conventional heating, where heat energy is applied to the outside surface and is conducted to the centre of food. Microwave ovens are thus 40% efficient as compared to 14% for standard electric ovens and 7% for gas ovens.

Mode of Action on Microorganisms

Microbial inactivation is achieved due to irreversible heat denaturation of enzymes, proteins, nucleic acids or other cellular constituents vital to cell metabolism and reproduction resulting in cell death. Cell membranes may also be damaged. Cellular metabolities and cofactors crucial for performing the cellular functions may also leak through damaged membranes.

Applications

Microwave has found its applications in many areas including

a) *Blanching* : Boiling + microwave blanching of potatoes can kill peroxidase enzyme in 4 to 5 min as compared to 15 min for boiling alone.

b) *Cooking* : Precooking bacon, meat, patties and poultry parts. Doughnuts are cooked without using oil in microwaves.

c) *Dehydration* : Dehydration of fruit juices and pulps using microwave ovens at 45°C and 6-8 torrs vacuum. It is advantageous to use microwave heating during falling rate period, when heat and mass transfer in conventional drying are least. It can also prevent case hardening at food surfaces. First commercial application of microwave energy was finish drying of potato chips.

d) *Pasteurization* : It takes 30-45 min to reach pasteurization temperature in thermal processing as compared to only 3-5 min with microwaves.

e) *Sterilization* : Metal package should not be used as metals are reflectors of microwaves. Most foods can be sterilized by microwaves.

f) *Tempering* : It is the process of raising the temperature of solidly frozen foods to a temperature just below the freezing point of water e.g. -4 to -2°C. Tempering is used in place of complete thawing in some of the packed frozen foods. It can be completed in minutes by using microwaves compared to 2-5 days in a conventional thawing room without unpacking of product.

g) *Baking* : Microwaves also find application in baking especially during the last phase when conventional heating takes longer time and results in excessive colour changes.

Factors Affecting Microwave Heating

a) *Starting temperature* : Higher the initial temperature, faster the microwave heating. Penetration of microwaves is easier and deeper into ice as compared to water but the absorption rate is higher for water than for ice and so is the efficacy of microwave heating.

b) *State of food* : Wet foods are heated more efficiently than dry foods.

c) *Homogeneity* : More homogenous the food, greater and more even is the absorption of microwaves and less time required for heating.

d) *Geometry* : Cubical or cuboidal slices may exhibit "edge over heating" and "corner heating".

e) *Quantity of food* : For every additional unit quantity of food, approximately one half of the time for one unit is added to the required heating time.

Dielectric Heating

It is based on the similar principle as microwave heating but operates at lower frequencies. Food is passed between capacitor plates and high frequency energy using an alternating electrostatic field is applied. This results in changing the orientation of dipoles as occurs in microwave processing. However, the thickness of the food cannot be more than the distance between the capacitor plates. This is an important limitation of dielecting heating.

Applications

a) Drying crispbread and biscuits

b) Thawing the blocks of frozen foods (e.g. egg, meat, fish, fruit juices)

c) Melting fats or chocolate.

Points to Remember

- √ Alpha, beat and gamma rays have < 100 nm wavelength
- √ Radiowaves have 30 m to 3 Km wavelength
- √ Microwaves are generated by magnetron at (2450 MHz or sometimes 896 MHz in Europe and 915 MHz in USA
- √ Magnetron is a device that converts electric energy at low frequencies into an electromagnetic field with centers of positive and negative charge that change direction billions of times per second. This causes the instantaneous penetration and heating of foods.
- √ Microorganisms die due to irreversible heat denaturation of enzymes, proteins, nucleic acids or other cellular constituents vital to cell metabolism and reproduction resulting is cell death.
- √ Dielectric heating operates using a similar principle to microwave heating but at lower frequencies.

5. Pulsed Electric Fields

Pulse electric fields (PEF) inactivate microorganisms in foods without affecting physical, chemical, sensory and nutritive characteristics of foods, as observed in traditional thermal processing methods. Application of pulse electric fields started in 1920's and 1930's when electricity was first used for pasteurization of milk in USA and the process was coined as "Electropure Process". Use of high voltage (AC/DC) electricity through electrodes, submerged in a media of microorganisms (waste water) during 1950's was studied (Electrohydraulic Treatment). Microbial inactivation was attributed to cavitation and wave impaction created by shock waves (upto pressure 1000 bars) that are produced by high intensity transient pressure pulses. PEF inactivation of microorganisms is based on electroporation of microbial cells by subjecting them to PEF of high intensity, that would cause destruction of microbial cells (Kumar *et al* 2000).

Theory

Various theories of microbial inactivation have been propounded to explain the actual mechanism of killing microorganisms by PEFs (Kumar *et al* 2000).

1. *Transmembrane potential (TMP)* : It is the difference in potential between bacterial cell and its external environment. TMP of cell = 1 volt.
   - When subjected to high intensity electric field i.e. TMP reaches 1 volt, the integrity of cell membrane is lost the cell is damaged.
   - *Breakdown TMP value* is the TMP at which the cell is affected/damaged.
   - Practically, TMP should be considerably higher than the critical value so that the cells could not rejuvenate.
2. *Dielectric Rupture* : This theory is based on the difference in dielectric constant between cell and foods (60-80) which causes cell inactivation.

When electric field is applied externally, it induces an electric potential (TMP) over the cell membrane, which in turn causes a charge separation in cell membrane. When TMP crosses a critical value repulsion between charged molecules forms pores in the bacterial cell membrane, increase cell permeability through which all the cell sap comes out and cell dies.

3. *Viscoelastic Effect* : The ability of a cell to withstand forces to which it is subjected to is referred as *viscoelasticity*.

   When a cell is subjected to PEF the electrocompressive forces will be counterpoised by viscoelastic restoring forces of cell membrane but to a certain level only. When electrocompressive force > viscoelastic force, the cell membrane ruptures and pores are formed through which cell sap flows out and bacteria dies. The size of pores may be directly correlated to the intensity of PEF applied.

4. *Electroporation* : Electroporation of cell membrane occurs due to denaturation of proteins or gelatinization of dipolar lipids in the cell membranes due to joule heating or electric modifications of -OH-, carbonyl, -SH- or amino groups. Thus, cell permeability is affected causing cell inactivation.

5. *Electro Mechanical Instability* : Application of high intensity PEF subjects cells to considerably higher electrical and mechanical forces (because of pulses), which causes cell inactivation, the time for cell lysis being reduced significantly if cell membrane is weak.

6. *Fluid Mosaic Arrangement* : When electric fields are applied, a dipolar reorientation in phospholipid structure of the orderly structured arrangement in cell membrane occurs, which leads to cell inactivation.

7. *Osmotic Swelling* : Cell inactivation is a result of osmotic imbalance across cell membrane induced by electroporation.

8. *Other Theories* : Some other theories discuss enzyme inactivation in a cell, oxidative reactions and

conformational change in tertiary structure of proteins promoting aggregation in cell membranes.

By and large, all the above theories are based on the formation of pores and rupture of cell membrane, ultimately leading to the death of microbial cells as the major reason for preservation of foods on application of PEF without generation of heat.

PEF Processes

PEF systems are designed for both batch process and continuous process. In static/batch systems food is held constant while in continuous system food is circulated by a pump into treatment chamber.

A higher degree of reduction in microbial counts indicates higher process efficiency and vice-versa. The various *Process Variables* such as PEF treatment time, PEF strength, pulse type, treatment temperature, number of pulses, type of system used, growth stage of microorganisms, osmotic pressure, electrical conductivity of medium/food and hurdles used if any, have a significant role in altering the process efficiency of PEF system.

- Deareation and bubble inhibition unit prior to treatment chamber should be provided to avoid dielectric breakdown causing a decrease in process efficiency.
- Do not let the food temperature increase to avoid electric resistance heating or ohmic heating.
- PEF energy should be high enough (short duration high voltage electric field pulses) for achieving satisfactory inactivation but not too high to initiate undesirable electrolytic reactions, ohmic heating and disintegration of food particles
- Impurities are separated before PEF processing as they may lead to dielectric breakdown.
- In continuous systems electrical and flow parameters are selected in such a manner that each unit volume of food is subjected to a specific degree of treatment achieving satisfactory degree of microbial inactivation.

Applications

Generally electric field strength of 12-35 KV/cm in short pulses of 1-100 μs is sufficient to produce a lethal effect on microorganisms in food. However, presence of hurdles may supplement the effect of applied PEF.

Critical electric field strength (CEFS) = 15 KV/cm eg. CEFS for *Lactobacillus brevis* and *Escherichia coli* are 13 KV/cm and 16 KV/cm respectively. Much higher CEFS is required for inactivation of asco and endospores.

PEF have found their commercial applications in

a) Pasteurization of liquid juices, soups, milk, orange, pineapple juice, sauces, etc.
b) Accelerated thawing
c) Decontamination of heat sensitive foods
d) Osmotic dehydration of carrots etc.

Points to Remember

- √ Pulse electric fields (PEF) inactivate microorganisms in foods without affecting physical, chemical, sensory and nutritive characteristics of foods, as observed in traditional thermal processing methods.
- √ When electricity was first used for pasteurization of milk in USA and the process was coined as "Electropure Process".
- √ PEF inactivation of microorganisms is based on electroporation of microbial cells by subjecting them PEF of high intensity, that would cause destruction of microbial cells
- √ Transmembrane potential (TMP) is the difference in potential between bacterial cell and its external environment. TMP of cell = 1 volt.
- √ Viscoelasticity is the ability of a cell to withstand forces to which it is subjected to
- √ Critical electric field strength (CEFS) = 15 KV/cm eg. CEFS for *Lactobacillus brevis* and *Escherichia coli* are 13 KV/cm and 16 KV/cm respectively. Much higher CEFS is required for inactivation of asco and endospores.

6. Magnetic Fields

In similarity with the pulsed electric fields, the magnetic fields can also be used for inactivation of microorganisms in various foods. The oscillating magnetic fields can inactivate microorganisms and pasteurize foods with an improvement in the quality and shelf life compared to conventional thermal pasteurization processes.

Theory

When a plastic package or bulk food is subjected an oscillating magnetic field, energy is coupled into the magneto-active parts of large biological molecules with several oscillations. A large number of magnetic dipoles when present in one molecule, enough energy can be transferred to the molecule to break a covalent bond. DNA or protein of the microorganisms could be broken by the treatment, hence destroying them or at least rendering them reproductively inactive. A metal package cannot be used in a magnetic field.

Following Theories were given to explain Inactivation Mechanism of Microorganisms (Pothakamury *et al.* 1993)

1. Weak Oscillating Magnetic Fields (OMFs) can loosen the bonds between ions and proteins. Many proteins vital to healthy metabolism contains ions. The effects of OMFs are pronounced at specific frequencies called as the cyclotron resonance frequency of ions. At 50 μ Tesla the resonance frequency of Na^+ and Ca^+ ions is 33.3 and 38.7 Hz respectively. At cyclotron resonance, energy is transferred from the magnetic field to the ions and also to the metabolic activities involving ions. This increases the velocity and ionic drift and therefore results in an increase in the net transport of Ca^{+2} ions across cell membrane. Increase in efflux of ions is frequency specific, while changes induced in the metabolic activities occur over a range of frequencies.

2. Static Magnetic Fields (SMFs) and Oscillating Magnetic Fields (OMFs) effect calcium ions bound in a calcium binding protein such as *Calmodulin*. These calcium ions continuously vibrate about an equilibrium position in the binding site of Calmodulin. On application of SMF to Calmodulin the plane of vibration rotates and results in loosening of the bonds between Calcium ions and the Calmodulin which may affect the metabolic activities of the cells.

Effective Dose

Magnetic field intensity between 5-50 Tesla, with short pulses of 25 μsec to 10m sec upto few minutes and frequency of 5-500 KHz are required for inactivation of microorganisms. The total exposure time is minimal and there is no significant rise in the temperature. *Streptococcus thermophilus* in milk, *Saccharomyces* in yoghurt and orange juice and unidentified spores in dough are reduced by several log cycles (2-3 log) after 1-10 pulses of 7.5-40 Tesla and 6-416 KHz magnetic fields, without any significant change in sensory quality.

Mode of Action

Oscillating magnetic fields couple energy into the magnetically active parts of the large critical molecules such as DNA. The amount of energy per oscillation coupled to one dipole in the DNA is 10^{-2} to 10^{-3} eV if magenetic field strength is 5-50 Tesla. Magnitude of the energy transferred becomes large with several oscillations and a large number of dipoles which may result in the breakdown of covalent bonds in the DNA molecule and inhibition of the growth of microorganisms.

The general, effect of the magnetic fields on microbial growth and reproduction can be classified as Inhibitory effect, stimulatory effect and no effect. The absence of any effects may be due to homogeneity of the fields. Inhibition is due to heterogenous fields. Magnetic fields also translocate free radicals such as $(C_6H_5)_3C^o$, ^oOH and ^oO to the region of greatest magnetic field intensity resulting in metabolic interruptions and inhibition of budding in yeasts cells subject to the condition that magnetic fields are perpendicular to the path of metabolism.

Applications

The use of magnetic fields have found their application in various processes such as pasteurization, sterilization and preservation of various foods like milk, yoghurt, fruit juices like orange juice, dough etc.

POINTS TO REMEMBER

- √ Magnetic field intensity between 5-50 Tesla, with short pulses of 25 μsec upto few minutes and frequency of 5-500 KHz are required for inactivation of microorganisms.
- √ Oscillating magnetic fields couples energy into the magnetically active parts of the large critical molecules such as DNA, resulting in the breakdown of covalent bonds in the DNA molecule and inhibition of the growth of microorganisms.
- √ The use of magnetic fields have found their application in various processes such as pasteurization, sterilization and preservation of various foods like milk, yoghurt, fruit juices like orange juice, dough etc.

7. Pulsed Light Treatment

Pulsed light treatment can be used for inactivation of microorganism at the surface of food and packaging materials and within transparent liquids, solids and gases. The light pulses used have duration of 10^{-6} to 10^{-1} sec, an energy density of 0.01 to 50 J per cm^2 and a wavelength spectrum between 170 to 2600 nm. Material for aseptic packaging or fluids such as air, water, or salt solutions, are treated with UV rich light pulses having $>$ 30 per cent light energy at wavelengths $<$ 300 nm. Application of intense light pulses on UV sensitive materials and foodstuffs transfers large amount of thermal energy to the surface of material and raise temperatures of thin superficial layer resulting in inactivation of surface microorganisms, enzymes, viruses. This process does not raise the interior temperature of the food product.

Theory

The anti-microbial effects of UV wavelengths contained in pulsed light spectrum are due to absorption by highly conjugated C=C double bond systems in proteins and nucleic acids which disrupts cellular metabolism. The lethal action of high intensity broad spectrum light is due to either photothermal (Visible spectrum) or photochemical (UV component) mechanisms (eg. formation of lethal thymin dimmers on the microbial DNA).

Equipment and Working

Pulsed light is produced as a result of magnifying power by accumulating electrical energy in an energy storage capacitor over relatively long time (fractions of a sec.) and release this stored energy to do work in much shorter time (millionth or thousandth's of a sec.). This results in very high power during duty cycle.

Generally broad spectrum white light flash containing wavelengths from 200 nm in UV to about 1 mm in near infra red is used for treatment of food and packaging material. Each pulsed light flash is about 20,000 times the intensity of sunlight at sea level. UV wavelengths are removed from the sunlight by the filtering effects of earth's atmosphere. The light energy at the surface is measured as *Fluence* or incident light energy per unit area in J/cm^2.

Applications

Pulsed light is used in

a) Waste water treatment
b) Sterilization of packages
c) Pasteurization of various liquid foods, fruit juices etc.
d) Preservation of baked goods, sea foods and meats (bread, cakes, pizza, shrimps etc.).

Comparison of Pulsed Electric Fields, Magnetic Fields and Pulsed Light Treatments

Pulsed Electric Field	Magnetic Field	Pulsed Light Treatment
1. *Mechanism of Action*		
Rupture of cell membrane caused due to difference in trans membrane potential and dielectric constants on application of HIPEFs	Loosening of bonds between ions and proteins present in the cell membrane which induces changes in metabolic activity of the microorganisms.	Combined photothermal and photochemical mechanisms (formation of lethal thymin dimmers in the DNA)
2. *Effective Strengths*		
Electric Field: 15-40 KV/cm Pulse Duration: Few - 200 μsec No. of Pulses: 1-20	Field Intensity: 5-50 Tesla Frequency: 5-500 KHz Exposure Time: 25 μsec	Pulsed Light wavelength 200-1000 nm Fluence: 0.5-2.0 J/cm^2
3. *Applications*		
1. Drying 2. Juice Recovery 3. Metabolite extraction from 4. Antimicrobial treatment of pumpable or liquid foods 5. Pasteurization and sterilization of liquid foods	1. Pasteurization and sterilization of various foods like milk, yoghurt, orange (fruit) juices, dough etc.	1. Shelf life extension and preservation of baked goods, Sea foods, meats, fruits and vegetables, water treatment.
4. *Advantages*		
1. No thermal degradation 2. Induces irreversible changes in the microbial membrane structures 3. Does not affect the visual quality of food. Retains more flavour and taste	1. Ensures no post-process contamination in packaged foods 2. Thermal energy input during exposure of foods is small 3. Minimal thermal denaturation of	1. 20,000 times the intensity of Sunlight 2. Brings about reduction in the microbial load by several log cycles in a small time. 3. No change in the nutritional and

contd...

contd...

than thermal processing	nutritional and organoleptic properties 4. Reduced energy requirements 5. Potential treatment for treatment of foods inside flexible packages	organoleptic properties of the food.
5. *Limitations*		
1. Mostly used for liquid foods 2. High pulsed energy applied may lead to ohmic heating 3. Liquids are required to be deaerated before HIPEF treatment	1. Magnetic fields may also have stimulatory effect or no effect on microorganisms if optimum field strengths and exposures are not given.	1. Mostly used for surface treatments 2. Sometimes UV portion of light is required to be filtered off as it may cause certain changes in taste of some foods.

Points to Remember

- √ Pulsed light treatment can be used for inactivation of microorganism at the surface of food and packaging materials and within transparent bulk liquids, solids and gases.
- √ The light pulses used have duration of 10^{-6} to 10^{-1} sec, an energy density of 0.01 to 50 J per cm^2 and a wavelength spectrum between 170 to 2600 m.
- √ Application of intense light pulses transfers large amount of thermal energy to the surface material and raise temperature of that thin superficial layer resulting in inactivation of surface microorganisms, enzymes, viruses.
- √ The anti-microbial effects are due to absorption of energy by highly conjugated C=C double bond systems in proteins and nucleic acids which disrupts cellular metabolism. The lethal action of high intensity broad spectrum light is due to either photothermal or photochemical mechanisms (eg. formation of lethal thymin dimmers on the microbial DNA).
- √ Pulsed light is used in waste water treatment, sterilization of packages, pasteurization of various liquid foods, fruit juices etc., preservation of baked goods, sea foods and meats (bread, cakes, pizza, shrimps etc.).

8. High Pressure Processing

Although the first reported use of high pressure processing of milk, fruits and fruit juices, meats etc. dates back to 1899 in West Virginia University, USA, the first commercial products produced by this technology were reported in 1990, and the products were jams of apple, kiwifruit, strawberry, raspberry, orange and grapefruit juices, fruit jellies, sauces, fruit yoghurts and salad dressings. These products sell at 3-4 times the cost of conventionally processed products due to their better quality, flavour, colour and texture.

Theory

For high pressure processing, packaged food is submerged in a liquid and high pressures upto 1000 MPa (10,000 bar) are applied. The liquid distributes the pressure instantly and uniformly throughout the food (isostatic) to kill the micro-organisms. High pressures cause collapse of the intracellular vacuoles and damage the cell wall and cytoplasmic membranes. Microorganisms in the log phase are more sensitive to high pressures (*Barosensitive*) than those in stationery, dormant and death phases. 350-400 MPa for 5-30 min cause about 10 folds reduction in vegetative cells of microorganisms. However, food enzymes vary in their barosensitivity. Some are inactivated even at few hundred MPa while the others (e.g. peroxidase in peas, and pectin methyl estrase in strawberries) can withstand as high as 1000 to 1200 MPa. High pressure also effect various types of chemical bonds such as ionic, hydrogen and hydrophobic bonds but does not effect the covalent bonds. Therefore, it can be used to kill microorganisms without significantly effecting the food molecules that contribute to the texture and flavour of food. Application of high pressures in pulses are more effective than static high pressures.

Applications

The high pressure processing has found its application in pasteurization, sterilization of a variety of plant and animal foods such as sauces, pickles, yoghurts, salad dressings, juices, james, jellies, fruit products, meat and vegetable products etc. Besides, the high pressure processing has also been used for the decontamination of heat sensitive food ingredients such as vitamins, flavours etc.

Merits

- The process can be operated at ambient or even chilled temperatures, therefore there is little heat damage to the quality attributes like nutrients, colour, flavour etc.
- Since the packaged food is subjected to high pressures, the process does not require use of chemical preservatives to achieve adequate shelf life of processed products.
- There is uniform application of high pressures to all parts of the food irrespective of its size or shape, while in conventional heat processing methods, the heat travels form the edges of food inwards through conduction or convection.
- Reduces the processing time, since there are no heating or cooling periods, only pressurization and de-pressurization cycles are there.
- Applicable to all types of food, whether solid or liquid.
- The risk of post-processing contamination is minimum.
- There is a greater choice in the type of container
- Little chance of any toxicity.

Limitations

- Expensive machinery
- Complex material handling
- Little effect on food enzyme activity.
- Little scope for continuous processing. Processing is to be done in batches.

POINTS TO REMEMBER

- √ For high pressure processing, packaged food is submerged in a liquid and high pressures upto 1000 MPa (10,000 bar) are applied.
- √ The liquid distributes the pressure instantly and uniformly throughout the food (isostatic)
- √ Microorganisms in the log phase are more sensitive to high pressures (*Barosensitive*) than stationery, dormant and death phases.
- √ 350 MPa for 30 min or 400 MPa for 5 min cause about 10 folds reduction in vegetative cells of bacteria, yeasts or moulds.
- √ The high pressure processing has found its application in pasteurization, sterilization of a variety of plant and animal foods such as sauces, pickles, yoghurts, salad dressings, fruit products, meat and vegetable products etc. Besides, the high pressure processing has also been used for the decontamination of heat sensitive food ingredients such as vitamins, flavours etc.
- √ The process can be operated at ambient or even chilled temperatures, therefore there is little heat damage to the quality attributes like nutrients, colour, flavour etc.
- √ Little chance of any toxicity.

9. Ultrasounds

Ultrasounds are the sound waves having frequencies above 16 KHz that cannot be detected by human ear. The low intensity ultrasounds which is less than 1 W cm^{-2} are used as a non-destructive analytical method to assess the composition, structure and flow rate of foods, while, the high intensity ultrasounds (10-1000 Wcm^{-2}) are used at higher frequencies (upto 2.5 MHz) to cause physical disruption of tissues, create emulsions, clean equipment or promote chemical reactions (eg. oxidation).

Theory

Ultrasounds produce rapid localized changes in pressure and temperature resulting in shear disruption, cavitation (i.e. creation of bubbles in liquid foods), thinning and weakening of cell membranes, localized heating and free radical production, leading to lethal effects on microorganisms (Fellows, 1988). The shearing and compression effects of ultrasounds which are produced on generation of force respectively in parallel and perpendicular directions to food, cause denaturation of proteins, reduced enzyme activity and rapidly changing pressure created by ultrasounds waves are effective in destroying microbial cells.

Applications

Ultrasonic waves have found use in tenderization of meat, inactivation of enzymes and destruction of microbial cells. A combined heat and ultrasounds treatment (Mano-Thermo-Sonication, MTS) is reported to be 6-30 times more lethal to the microorganisms as compared to the sole heat treatment at same temperatures. MTS is more effective against yeasts as compared to that against the bacterial cells. The death occurs in logarithmic cycles. Ultrasounds also assist drying and diffusion (acoustic drying), in some foods like gelatin, yeasts and orange powder, where the rates of drying increase by 2-3 times, possibly due to creation of microscopic channels in the solid foods by oscillating compression waves and by changing pressure gradient at the air/ liquid interface resulting in increased rates of evaporation.

Merits

- Reduce heat resistance
- Effective for inactivation of enzymes
- Can be combined with other processing methods like drying, heat processing etc.

Limitations

Resistance of microorganisms and enzymes to ultrasounds is so high that the intensity of treatment may produce adverse changes to the texture and other physical properties of food and substantially reduce its sensory quality.

POINTS TO REMEMBER

- √ Ultrasonics are the sound waves having frequencies above 16 KHz that cannot be detected by human ear.
- √ Ultrasounds produces rapid localized changes in pressure and temperature that causes shear disruption, cavitation (creation of bubbles in liquid foods), thinning of cell membranes, localized heating and free radical production, which have lethal effects on microorganisms
- √ Ultrasonic waves have found use in tenderization of meat, inactivation of enzymes and destruction of microbial cells.
- √ A combined heat and ultrasounds treatment (Mano-Thermo-Sonication, MTS) is reported to be 6-30 times more lethal to the microorganisms as compared to the sole heat treatment at same temperatures.
- √ Resistance of microorganisms and enzymes to ultrasounds is so high that the intensity of treatment would produce adverse changes to the texture and other physical properties of food and other substantially reduce its sensory quality.

10. Linear Induction Electron Accelerator (LIEA)

LIEA is an electrically driven source of ionizing radiation which overcomes the problems associated with the radionuclides (Co-60 and Cs-137) used in irradiation. In similarity to the radionuclides, the LIEA generates (i) ionizing radiations (radiations that are capable of producing ions) and (ii) excited atoms or molecules which induce chemical changes in the food and in the biological cells leading to inactivation of microorganisms. But, the radiations are generated by electrically driven source instead of radionuclides.

Theory and Equipment

In contrast with the radionuclides Co-60 and Cs-137, which emit γ-radiations of discrete energy (i.e. 1.17 MeV and 1.33 MeV for Co-60 and 0.67 MeV for Cs-137), the LIEA generates broad spectrum ionizing radiations. These are produced by collision of accelerated electron beam and a heavy metal like Pb-Convertor Plate. This plate converts the electron beam into X-rays with broad band of photon energy spectrum. The LIEA is different from other electrically driven sources due to its high beam power and reliability in operation.

Advantages

- It is an electrically driven radiation source that can be switched off when no longer in use.
- Does not require permanent massive concrete shielding as in case of irradiation.
- There is possibility of controlling the directions of the electrically produced radiations.
- Possibility of shaping the geometry of radiations to accommodate different sizes.
- Can deliver dose rates many orders of magnitude higher than possible with Co-60 sources.
- Ultra short high intensity irradiation treatments can be applied, which result in higher local radical concentrations, favouring radical-radical recombinations. This reduces the diffusion of radical species which are thought to be responsible for undesirable effects of irradiations on food quality (Mertens and Knorr, 1992).

Points to Remember

√ LIEA stands for linear induction electron accelerator

√ LIEA generates broad spectrum ionizing radiations by targeting the accelerated electron beam to collide with a heavy metal like Pb-Convertor Plate. This plate converts the electron beam into X-rays with broad band of photon energy spectrum.

√ Does not require permanent massive concrete shielding as in case of irradiation.

11. Minimal Processing and Hurdle Technology

Minimal Processing : Technology of preservation of food by minimum application of heat (thermal minimal processing) and using alternative non-thermal methods (PEF, magnetic fields, high pressure, ultrasounds, ozone) so as to obtain a shelf stable product having fresh-like qualities and minimum losses in nutritional, sensory and functional properties.

Hurdle technology : It is also called combined methods technology. By hurdle technology or combined methods technology, the shelf life and microbial stability of different types of foods is extended by using several factors (hurdles) in combination, none of which, individually, is able to check the spoilage completely and is effectively lethal for microorganisms.

Hurdle technology is a new concept developed for ensuring safe, stable, nutritious, tasty, and economical food by employing an intelligent simultaneous or sequential combination of different preservation factors or techniques (hurdles) to achieve multi-target, mild but reliable preservation effects.

Hurdles

Many hurdles have been identified so far, but most of the work refers to meat products. However, its application in bakery products, fish, dairy products, mild processing of fruits and vegetables is recent and gaining popularity.

Some of the important hurdles used are :

- Low water activity, drying and dehydration
- Low temperature, refrigeration, freezing
- Modified atmospheres and gas packaging
- Low pH, high acidity
- Chemical preservatives (sulphites, benzoates, nitrates, sorbates etc.)

- Application of mild heat
- Bacterocins,
- Ultra-high pressure treatment
- Edible coatings
- Redox potential
- Mano-thermo-sonication etc.

Homeostasis

It is the constant tendency of microorganisms to maintain a stable, uniform and balanced internal environment. Hurdles disturb one or more of the homeostasis mechanisms, thus prevents microbial multiplication and cause them to remain inactive or even die. Disturbing several homeostasis mechanisms simultaneously increases efficacy of the hurdle technology.

Merits

- Better retention of nutritional and sensory quality of food due to minimal heat application
- More effective synergistic effect of hurdles
- Better consumer preference
- Fresh or fresh like products

Limitations

- Individual hurdles do not provide effective preservation so the extent/intensity of use of hurdles should be perfectly ensured.
- Proper packaging is very important
- Requires more skill than that for traditional heat preservation

Points to Remember

√ Minimal Processing is the technology of preservation of food by minimum application of heat (thermal minimal processing) and using alternative non-thermal methods (PEF, magnetic fields, high pressure, ultrasounds, ozone) so as to obtain a shelf stable product having fresh-like qualities and minimum losses in nutritional and functional properties.

√ *Hurdle technology* : It is also called as combined methods technology. By hurdle technology or combined methods technology, the shelf life and microbial stability is extended by using several factors (hurdles) in combination, none of which, individually, is able to check the spoilage completely and would be totally lethal for microorganisms.

√ Bacterocins, ultra-high pressure treatment, edible coatings, redox potential, mano-thermo-sonication etc. are new hurdles

√ Individual hurdles do not provide effective preservation so the extent/intensity of use of hurdles should be ensured.

12. Functional Foods and Nutraceuticals

Functional Food

This is similar to coventional food and is consumed as a part of normal diet but is known/proven to have physiological benefits and capacity to reduce risk of chronic diseases beyond basic nutritional functions.

Nutraceuticals

Nutraceuticals are foods or a part/constituent of a food with specific medical or health benefits, for prevention, treatment or cure of a disease. A neutraceutical can also be defined as "a diet supplement that provides a concentrated form of a bio-active compounds, of food, in a non-food matrix to enhance health and prevent diseases."

Example : You consume more of Vitamin A, carotene, green and yellow vegetables there would be lesser risk of cancer. This can be associated to antioxidant protection offered by these constituents in foods. Vitamins C, E and beta carotene are considered to be antioxidant vitamins.

Antioxidants

Free radicals are produced in our body as a normal part of life. These free radicals are very reactive and readily oxidize the fats present in cell membranes. Oxidative degradation causes cell damage and plays a role in aging and disease. Antioxidants prevent oxidation of fats and prevent ageing and diseases.

Designer Foods

These are specialized processed foods supplemented with the ingredients rich in cancer/disease preventing substances. Some of the foods are garlic, cabbage, licorice, soybeans, ginger, and umbelliferous vegetables (carrots, celery, parsnips). So designer foods can also be understood as a type of functional food.

Consider the Following Types of Foods

1. Raw vegetables and fruits- carrots and mango contain β-carotene.
2. Processed foods without added ingredients-oat bran, wheat porridge
3. Processed foods with added ingredients-Ca enriched juice, Fe fortified flour
4. Genetically modified foods-tomato having higher lycopene
5. Purified preparations of isolated active ingredients of various plants sold as tablets or capsules etc.

Among the above five types first four are functional foods, 3rd specifically may be referred to as designer food, and fifth may be neutraceutical. However, functional foods, neutraceuticals, designer foods, pharma foods are all synonyms for food that can prevent or treat diseases (Goldberg 1994).

Some Active Compounds

Compouns	Plant in which found	Properties	Use
Butyl phthalide	Celery	Provides distinctive taste and smell	Protection against cancer, high blood pressure and high cholesterol levels
Calcium pectate	Fruits and vegetables, apples, carrots, cabbages, onions	Responsible for crispness in fruits and vegetables	Potent cholesterol-lowering properties
Capsaicin	Chilli, Pepper	Mouth burning, eye watering, and breath-taking effects	Digestive aid, painkiller, potential cancer-fighting compound
Cathecin Hydrate	Tea (*Camellia sinensis)*	A tannin derivative that gives tea its astringency	Cancer-fighting properties have been attributed to tea
Coumarin	Tonka bean, lavender, sweet clover grass, licorice, strawberries, apricots, cherries, and cinnamon	A blood thinning agent that acts as a rat poison	Warfarin (a coumarin derivative) is the most commonly used oral anticoagulant medication. Anti-fungicidal, and anti-tumor activities.
Ellagic acid	Strawberries and raspberries (highest qty), blackberries, cranberries, walnuts, and pecans	A natural phytochemical pesticide in many fruit plants	This phytochemical fights cancer in humans
Heliotropin	*Cinamomum petrophilum* and *Sassafras albidum*	Relaxing properties in aromatherapy	Manufacture of perfumes and soaps, Used to manufacture designer drugs, such as Ecstasy and methylenedioxyamphetamine (MDA)
Lutein	Spinach, kale, collared greens, romaine lettuce, leeks and peas	Fat-soluble antioxidant carotenoid biochemical	Lutein and related carotenoids are now believed to protect against development of cataracts
Lycopene	Tomatoes, watermelon, grapefruit, guava, rosehip, and red chilies	Responsible for the red, orange, and yellow colour in fruits and vegetables	Protects against heart disease, certain cancers, and a multitude of other disorders
Saponin	Yucca and Quillaja	Saponins are natural surfactants, or detergents	Antimicrobial, cholesterol-lowering, and anticancer phytochemicals. Some saponins, such as digitalis, are also used as heart medications
Sulforaphane	Broccoli	Sulforaphane is a naturally occurring sulfur-containing isothiocyanate derivative	Helps to mobilize the human body's natural cancer-fighting resources and reduces the risk of developing cancer
Zeaxanthin	Spinach	Zeaxanthin is a yellow-coloured lipid-soluble xanthophyll, which is also an oxidized hydroxy derivative of *beta*-carotene	Protect against development of cataracts

There are 3 conditions that a functional food must satisfy

- They are foods not capsules, tablets or powder which are derived from naturally occurring ingredients
- Should be consumed in daily diet as full or part
- They generally perform a particular function of prevention or cure of diseases when ingested.

Phytochemicals

These are chemicals that are produced by plants. Here, the term refers to only those plant chemicals that may have health-related effects but are not considered essential nutrients (proteins, carbohydrates, fats, minerals, and vitamins). Many plant compounds are disease fighters in the body, boosting production or activities of enzymes, which then act by blocking carcinogens, suppressing malignant cells, or interfering with the processes that can cause heart disease and stroke and various other diseases. Some of such phytochemicals have been detailed ahead.

a. Allyl Sulfur Compounds

Found in members of the Alliaceae family (onions, garlic, and leeks). Allyl sulfur compounds are cysteine sulfoxide derivatives such as allicin and alliin.

b. Ajoene

Ajoene (a component of allicin) is a naturally occurring "nutriceutical" that reduces the risk of stroke and heart disease. It is an unsaturated sulfoxide disulfide, responsible for garlic's anticoagulant (blood thinner) properties and gives garlic its characteristic odour and flavor. Ajoene also has been found to have effective antimicrobial properties, inhibiting the growth of both bacteria and fungi and use in treatment of blood clotting disorders.

c. Glucosinolates

Glucosinolates are glucosides of sulfur-containing amino acid derivatives found mainly in *Brassica* members.

d. Polyphenols and Flavonoids

These compounds have no known role in human nutrition (non-nutrients), but have properties like antioxidant, antimutagenic, anti-oestrogenic, anti-carcinogenic and anti-inflammatory effects that might potentially be beneficial for disease prevention. High green tea consumption in Japan and moderate red wine consumption in the French population is considered to be beneficial for heart disease and cancer, due to specific polyphenols.

Flavonoids known as vitamin P (for permeability) have effects on the permeability of blood capillaries are also dietary non essentials. Flavonoids include proanthocyanidins, quercetin, resveratrol and epicatechin, found in chocolate, tea, and wine.

e. Phytoestrogens

Many plant compounds (glycosides) having weak estrogenic/ antiestrogenic actions, are collectively known as phytoestrogens. They have antibacterial or antifungal actions in the plant. High consumption of legumes i.e. soya beans, (rich sources of phytoestrogens), results in lower incidence of breast, and uterine cancer and osteoporosis.

f. Genistein

Found in soy products, genistein is an isoflavone derivative, a strong inhibitor of cancer.

Some Commercial Nutraceutical Products

Trade name of product	Contents	Functions	Manufacturer
GRD	Proteins, vitamins, minerals, carbohydrates	Nutritional supplement	Zydus Cadila Ltd., Ahmedabad, India
Proteinex	Proteins, vitamins, minerals, carbohydrates	Nutritional/ Protein supplement	Pfizer Ltd., Mumbai, India
Calcirol D-3	Calcium and vitamins	Nutritional/ Ca supplement	Cadila healthcare Ltd. Ahmedabad, India
Threptin Diskettes	Proteins and vitamins	Nutritional/ Protein supplement	Raptakos, Brett & Co. Ltd. Mumbai, India
Weight Smart	Vitamins and trace elements	Nutritional supplement	Bayer corporation, Morristown, NL, USA

POINTS TO REMEMBER

- √ *Nutraceuticals* are foods or a part/constituent of a food with specific medical or health benefits, for prevention, treatment or cure of a disease.
- √ Anti-oxidants prevent oxidation of fats and prevent ageing and diseases.
- √ Designer foods are processed foods supplemented with food ingredients rich in cancer/disease preventing substances. Some of the foods are garlic, cabbage, licorice, soybeans, ginger, and umbelliferae (carrots, celery, parsnips).
- √ Phytochemicals are chemicals that are produced by plants but not considered essential nutrients (proteins, carbohydrates, fats, minerals, and vitamins).
- √ Allyl Sulfur Compounds are found in garlic, onions, and leeks
- √ Ajoene is a naturally occurring "nutriceutical" that reduces the risk of stroke and heart disease.
- √ Flavonoids also were known as vitamin P (for permeability), because they were shown to have effects on the permeability of blood capillaries but were not dietary essentials.

13. Genetically Modified Foods

Genetically Modified (GM) Foods

GM foods are the foods/food products that have their DNA directly altered through genetic engineering (i.e. removing one gene from one organism (plant, animal or microbe) and transferring it to another, without affecting any other aspect of plant or animal). GM foods are known to contain new proteins produced by recombinant DNA technique and are also called transgenic or biotech foods (Rao 2004). By this technology desirable traits can be introduced in the crops without affecting other properties.

- GM foods first appeared in the market in the early 1990s. Common GM foods are derived from plants i.e. soybean, corn, canola, and cotton seed oil.

- The first commercially grown genetically modified whole food crop was the tomato (called Flavr Savr), which was made more resistant to rotting by Californian company Calgene. A variant of the Flavr Savr was used to produce tomato paste.

Desirable Traits to be Introduced in GM Foods (Rao, 2004)

a. Insect - pest resistance : Bt cotton, pest resistant brinjal, cabbage, tomato, herbicide tolerance, viral resistance (tobacco, papaya, peanut, tomato), fungus resistance (potato)
b. Improved quality : Tomatoes with delayed ripening, improved shelf life of fruits and vegetables, commodity with better flavour and colour
c. Adaptation to stress conditions : Draught and high temperature resistance, high salt resistance, adaptability to low fertility areas
d. Nutritional enrichment : Reduced saturated fatty acids in oil crops, low cholesterol eggs, soya with high methionine, beta carotene rich rice and sorghum
e. Functional foods : edible vaccines

Issues

a. Potential toxicity
b. Allergenicity form parent gene and from new protein
c. Cross pollination form GM to non GM crops
d. Ethical and religious concerns involving special animals and transfer of genes from and to them
e. Loss of biodiversity.

Labelling of GM Foods

- Mandatory to label the foods if it is genetically modified
- Processed foods prepared from GM plant/ sources should also be labeled

- The rules and procedures for handling GM foods in India are being monitored by IBSC (Institutional Bio-Safety Committee) and Review Committee on Genetic Manipulation (RCGM)
- COT i.e Committee on Toxicology is the nodal agency for studying and approval of GM crops or GM foods in UK
- Codex Alimentarious Commission - Body responsible for development of international standards for food safety and consumer protection
- USDFA regulates development of GM foods in USA.

Points to Remember

√ GM foods are the foods/food products that have their DNA directly altered through genetic engineering (i.e. removing one gene from one organism (plant, animal or microbe) and transferring it to another, without affecting any other aspect of plant or animal).

√ GM foods are known to contain new proteins produced by recombinant DNA technique and are also called transgenic or biotech foods

√ The first commercially grown genetically modified whole food crop was the tomato (called Flavr Savr), which was made more resistant to rotting by Californian company Calgene. A variant of the Flavr Savr was used to produce tomato paste.

√ GM foods can be toxic and allergent.

√ The rules and procedures for handling GM foods in India are being monitored by IBSC (Institutional Bio-Safety Committee) and Review Committee on Genetic Manipulation (RCGM)

√ COT i.e Committee on Toxicology is the nodal agency for studying and approval of GM crops or GM foods in UK

√ Codex Alimentarious Commission - Body responsible for development of international standards for food safety and consumer protection

√ USDFA regulates development of GM foods in USA.

14. GAP and GMP

Globalgap Certification for Good Agricultural Practices

Global expansion in food trade and World Trade Organization (WTO) aims to open up trade between countries. Sanitary and phyto-sanitary (SPS) issues are important in global trade and have become the most important potential Technical Barriers to Trade (TBT). Pests or pathogens may exist in one country but not in other, thus ultimately resulting in restrictive TBT.

For future trade between developed and developing countries, the food safety shall be the most important minimum requirement. Increasing outbreaks of food-borne diseases necessitates establishing minimum food safety standards. Global trade in fresh and processed food products would require compliance to some kind of food safety assurance system.

Safe food must be ensured in both developed and developing countries by enforcing appropriate legislation. In European Union, the importance of food safety was emphasized by the outbreaks of Mad Cow disease, Foot and Mouth disease as well as traditional concerns with environmental pollution, particularly pesticides and the issues surrounding Genetically Modified Organisms (GMO). However, in USA the main focus of concern is the outbreaks of food borne diseases associated with the consumption of fresh or processed food.

The increased concern of people all through out the world, primarily in developed countries, for eating healthy, safe and fresh food, lead to the establishment of the Good Agricultural Practices (GAP) policies and surveillance systems.

Some of the systems that are adopted to ensure safe food production are

a. Good Agricultural Practices (GAP)

b. Good Manufacturing Practices (GMP)

c. Good Handling Practices
d. Hazard Analysis Critical Control Points (HACCP)
e. Good Hygiene Practices (GHP) etc.
f. Globalgap : Within the European community. Apart from Germany and France, most other countries within the EU support this system, as do the major retailers, which consider it the minimum standard for food trade.

Efforts are being made to harmonize these global standards all over the world but for the time being, retailers have their own set of requirements that producers should adhere to.

EUREPGAP

GLOBALGAP started as a retailer initiative in 1997 and was established by the Euro-Retailer Produce Working Group (EUREP) with the aim of setting standard and procedures for the development of GAP.

Objectives of EUREPGAP

The main objective of Globalgap is, to lead the system to an EN 45011- based accredited certification system, referring to the cope of "Globalgap Fruits and Vegetables".

- Certification to Globalgap has become mandatory as from March 2003 for farms growing produce for export to Europe.
- Globalgap certification will become a minimum requirement for entry into the EU market.
- Additional retailer requirements will still have to be met.
- Globalgap only covers produce up to the farm gate and thereafter other systems such as GMP, HACCP etc will

become essential. All food industries must also implement GMP and GHP, both of which are prerequisite programs for HACCP.

- Certification to GLOBALGAP will result in additional costs to growers, there will be numerous benefits also.

Benefits

- More motivated farm workers due to improved facilities, training and better working conditions with a subsequent increase in living standards.
- Result in better productivity and outputs to the ultimate benefit for the grower.
- More environmentally sound farming practices
- More judicious use of chemicals
- Cost benefit to the grower due to better management practices enforced by the standard.

List of GLOBALGAP Certification Agencies in India

Name of the Certification Agency	Address & Contact Number	Contact Person & Email Address
Control Union Certifications	Summer Ville, 8th Floor, 33rd -14th Road Junction, Khar (W), Mumbai -400 052, India Tel: +91 -22- 67255390/91/92/93 Fax: +91 -22- 67255394/95 Website: www.controlunion.com	Mr. Sanjay Sailas, Sr. Inspector Email: sailas@controlunion.com cuc@controlunion.in, ssl@controlunion.in
ECOCERT India	Sector - 3, S-6/3 &4, Gut No. 102, Hindustan Awas, Walmi - Waluj Road, Nakshatrawadi - 431 002.	Dr. Selvam Daniel Country Representative and Director Certification Operations-India Email: office.india@ecocert.com

contd...

	Aurangabad , Maharashtra State, India. Telefax: +91 -240- 2377120, 2376949. Website: www.ecocert.in	ecocert@sancharnet.in
EUROCERT INDIA	Plot No 372, Phase I, Industrial Area 134113 Panchkula Tel: 172580467, 572900 Fax: 172 569849 Website: www.eurocert1.com	Ravinder Kakkar agri@iclcertifications.com
FoodCert India (p) Ltd	3-6-157 Himayatnagar, 4th floor Victory Vihar Appartments 500 029 Hyderabad Tel: + 91 40 66256146, 23221393 Fax: + 91 40 66256145 www.foodcertindia.com	Srihari Kotela srihari@foodcertindia.com
IMO CONTROL (Pvt.) Ltd.	No. 26, 17th Main, HAL A II Stage 560008 Bangalore Tel: +91 80 -52 01 546 Fax: +91 80 -52 72 185 www.imocontrol.net	Umesh Chandrasekhar imoind@vsnl.com
INDOCERT	Head Office: Thottumgham P.O Aluva, Emakulam Dist. Kerala - 683 105 Tel: 0484 262943, 2630908 Liaison Office: M/s. Indocert , Flat No. 3, Saket Aparments , Manekshanagar Near Kathe Galli , Dwarka, Nashik Maharashtra -422011 Ph:02532502213	Mr. Mathew Sebastian, Executive Director info@indocert.org

contd...

SGS India Pvt. Ltd.	2nd Floor, Ridhi-Sidhi App., Near Agarwal Petrol Pump, Sanjivani Nagar Ozar (mig), tal: Niphad Nashik-422206 Tel: +91-2550329887 Fax: +91-2550274913 www.sgs.com	Mr. Swapnil kadam Email: Swapnil_Kadam@sgs.com nasik_sgs@sgs.com
TUV SUD South Asia	Off Saki Vihar Road, Saki naka, Andheri (E), Mumbai - 400 072 Tel:+91-22-3082 3082/9797 Fax: +91-22-3082 9595 www.tuv-sud.in	Mr. Mahesh Deshpande Email: Mahesh.Deshpande@tuv.sud.in

www.msamb.com/english/export/globalgap.htm

Quality Standards

The quality standards that govern the food industry include;

- Prevention of Food Adulteration Act, 1954 (PFA), Vegetables Product Order, 1967 (VPO), Food Products Order, 1955 (FPO), Meat and Food Products Order, 1973 (MFPO), Meat and Meat Product Order, 1992 (MMPO). Agri-produce (Grading and Marketing) Act, 1937, Bureau of Indian Standard (BIS), Export (Quality Control and Inspection Act) (1963)
- There are few standards for unprocessed/raw food materials. Good Manufacturing Practices (GMP), Good Hygiene Practices (GHP), Hazard Analysis Critical Control Point (HACCP), Codex, International Standards Organisation (ISO) 9000 are some other certification procedures applicable to processing of foods and products.

- Codex covers 234 foods, 32 functional groups (additives), 36 flavouring agents, 6 labeling standards, 49 codes of practices, and 41 guidelines.
- BIS has about 700 Indian Standards for use in the area of agricultural produce and value added products.
- The key issues addressed by these standards include preventing adulteration, regulate hygienic conditions, inform consumers about the product, manufacturer etc., provide product specifications as well as export specifications
- An integrated food law viz., Food Safety and Standards Act 2005 has come into existence that consolidates most of the food related laws in India to meet the international standards.

Points to Remember

√ The increased concern of people all through out the world, primarily in developed countries, for eating healthy, safe and fresh food, lead to the establishment of the Good Agricultural Practices (GAP) policies and surveillance systems.

√ Good Manufacturing Practices (GMP)

√ Hazard Analysis Critical Control Points (HACCP)

√ Good Hygiene Practices etc.

√ GLOBALGAP : Within the European community. Apart from Germany and France, most other countries within the EU support this system, as do the major retailers, which consider it the minimum standard for food trade.

√ Globalgap started as a retailer initiative in 1997 and was established by the Euro-Retailer Produce Working Group (EUREP) with the aim of setting standard and procedures for the development of GAP.

15. HACCP

HACCP (Hazard Analysis and Critical Control Point) is a control system designed to identify and prevent or minimize different kinds of hazards of microbial / biological, chemical or physical origin, in the process of food production. It emphasizes upon various steps designed to prevent problems or hazards before they actually occur and to correct them as soon as they are detected.

Hazard

It refers to a potential undesirable and unacceptable danger of (i) biological origin (bacteria, viruses, protozoa, moulds, parasites etc.), (ii) chemical origin (toxins, pesticides, heavy metals, preservatives, colours and other additives etc.) or (iii) physical origin (fine glass pieces, metal fragments, stones, wood splinters etc.). Therefore, a food safety hazard is any biological, chemical, or physical cause that may make the food unsafe for human consumption.

Seven Principles / Components of HACCP

1. Hazard Analysis: *Identify the potential hazards, assess their extent of severity and the probability of occurrence of hazards.*
2. Critical Control Points: *Determine the critical control points (CCPs).* CCP is a point, stage, raw material, practice, formulation, process, step, or procedure in a food production process at which we can apply control (measures) and the hazard can be prevented, eliminated, or reduced to an acceptable level.
3. Critical limits for each CCP: *Determine the critical limits for each CCP.* Critical limits are the maximum or minimum values that will make a demarcation between the acceptability or unacceptability. Crossing any critical limit may make the stage, raw material, practice, formulation, process, step, or procedure, unacceptable in safe food production process.

4. Monitoring: *Monitor the manufacturing process continuously.* It is necessary to check the effectiveness of a control and ensure that the process is under control (within acceptable limits) at each CCP.
5. Corrective actions: *Take prompt corrective actions.* When monitoring indicates chances of any deviation from the established critical limits, prompt corrective actions are taken to ensure that no product injurious to health or otherwise adulterated as a result of the deviation is manufactured or enters the market.
6. Record Keeping: *Maintain the HACCP documents.* The documents including hazard analysis, written HACCP plan, monitoring of critical control points, critical limits, verification activities and handling of processing deviations is necessary for validation, verification, review, auditing, improvement etc.
7. Validation and Verification: *Establish procedures for ensuring that HACCP system is working as intended.* "Validation" of the HACCP plans ensures that the production plants are successful in ensuring the production of safe products. "Verification" is done to check overall effectiveness of the HACCP system and ensures that the HACCP plan is adequate, and working as intended. Verification procedures may include review of HACCP plans, CCP records, critical limits and microbial sampling and analysis.

WhyHACCP?

1. HACCP is necessary to ensure safe manufacture of various food items.
2. India a signatory of Sanitary and Phytosanitary (SPS) agreement. Therefore, we have to adopt standard guidelines and recommendations issued by FAO/WHO Codex Alimentarius Commission, for international trade, which also includes adoption of HACCP.

3. HACCP is mandatory for export from certain sectors of food industries to some countries.

Applications

HACCP has applications in various industries including juice, seafood, meat, poultry, cosmetics, pharmaceutical and various processed food industries.

Conception

HACCP was conceived in 1960s when the US National Aeronautics and Space Administration (NASA) asked "Pillsbury" to design and manufacture the first foods for space flights. Since then, HACCP is recognized at international level as a logical tool for adapting conventional inspection methods to a modern, science-based, food safety system. HACCP allow both industry and government to allocate their resources efficiently in establishing and auditing safe food production practices.

Points to Remember

- √ HACCP stands for Hazard Analysis Critical Control Point
- √ This system is designed to identify and prevent or minimize different kinds of hazards of biological / microbial, chemical or physical origin
- √ Hazard refers to a potential undesirable and unacceptable danger of biological origin (bacteria, viruses, protozoa, moulds, parasites etc.), chemical origin (toxins, pesticides, heavy metals, preservatives, colours and other additives etc.) or physical origin (fine glass pieces, metal fragments, stones, wood splinters etc.).
- √ There are seven principles or components of HACCP
- √ Critical Control Point (CCP) is a point, stage, raw material, practice, formulation, process, step, or procedure in a food production process at which we can apply control (measures) and the hazard can be prevented, eliminated, or reduced to an acceptable level.

- √ HACCP is mandatory for export from certain sectors of food industries to some countries.
- √ HACCP has applications in various industries including juice, seafood, meat, poultry, cosmetics, pharmaceutical and various processed food industries.

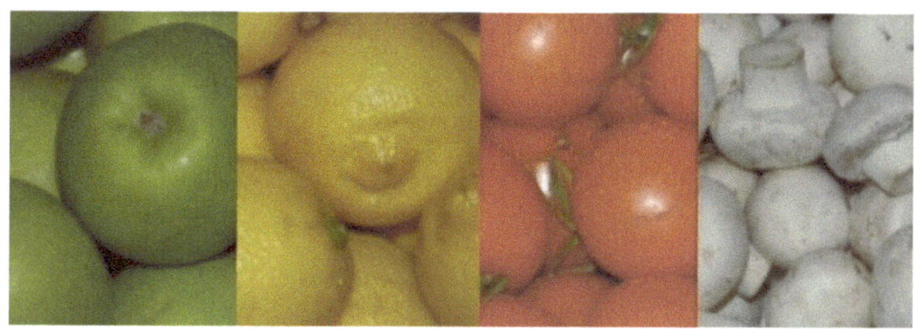

PART IV
Objective Questions

20

Evaluate Yourself

1. Multiple Choice Questions

I. Tick the Correct Answer

1. .... is used for extraction of essential oils
   a. Lemon — b. Olive
   c. Almond — d. Pineapple

2. Lye peeling is done by ....
   a. 1% NaOH + heat — b. 5% NaOH + heat
   c. 7% NaOH + heat — d. 0.1% NaOH + heat

3. .... is a fermented fruit beverage
   a. Cider — b. Syrup
   c. RTS — d. Ginger ale

4. .... is checked by adding preservatives
   a. Enzymes — b. Micro organisms
   c. Colour deterioration — d. All

5. .... is suitable preservative for red/ blue coloured products
   a. KMS b. Sodium Benzoate
   c. Both d. None

6. .... is suitable preservative for lemon juice
   a. KMS b. Sodium Benzoate
   c. Both d. None

7. KMS produces .... gas
   a. Sulphur dioxide b. Carbon dioxide
   c. Potassium dioxide d. Magnessium dioxide

8. Maximum permissible limit of use of SO_2 in Squash is ....
   a. 350 ppm b. 250 ppm
   c. 600 ppm d. 450 ppm

9. Maximum permissible limit of use of Sodium Benzoate in Squash and appetizer is ....
   a. 350 ppm b. 600 ppm
   c. 100 ppm d. 400 ppm

10. .... is known as vinegar bacteria
    a. Lactobacillus b. Bacillus
    c. Acetobacter d. Enterobacter

11. T stage is used as maturity index in ....
    a. Mango b. Apple
    c. Pineapple d. Cashewnut

12. Methionine is a precursor of ....
    a. Protein b. ethylene
    c. Vitamin d. IAA

13. Guava is a rich source of ....
    a. Vitamin A b. Vitamin C
    c. Vitamin D d. Vitamin E

14. Squash contains .... % fruit part
    a. 10
    b. 15
    c. 20
    d. 25

15. .... is also called spiced squash
    a. Beverage
    b. Appetizer
    c. Cordial
    d. Crush

16. Vase solutions contain .... concentration of sugar as/than pulsing solutions
    a. High
    b. Equal
    c. Low
    d. Nil

17. Glycolysis is also known as ....
    a. Kreb cycle
    b. EMP pathway
    c. NADH cycle
    d. None

18. Predominant acid in apple is ....
    a. Citric acid
    b. Ascorbic acid
    c. Malic acid
    d. Acetic acid

19. Dried grapes are called as ....
    a. Resin
    b. Rosin
    c. Raisin
    d. Resign

20. Most fruits are processed satisfactorily at ....
    a. 87° C
    b. 100° C
    c. 101° C
    d. 121° C

21. Most vegetables are processed satisfactorily at ....
    a. 87° C
    b. 100° C
    c. 101° C
    d. 121° C

22. Blanching is done to ....
    a. Partially kill the micro-organisms
    b. Completely kill all the microorganisms
    c. Inactivate enzymes
    d. Prevent fermentation

23. Pasteurization is done at ....
    a. 100° C
    b. > 100° C
    c. < 100° C
    d. 150° C

24. UHT processing is done at 130-150° C for ....
    a. 2-3 min b. 2-3 sec
    c. 2-3 hrs d. 30 min.

25. Osmotic dehydration of fruits is done in ....
    a. Sugar solution b. Salt solution
    c. Caustic soda solution d. KMS solution

26. Which of the following is called as protective foods
    a. Fruits b. Vegetables
    c. Fruits and Vegetables d. Milk

27. Fruits and vegetables are called protective foods because they contain ....
    a. Proteins b. Carbohydrates
    c. Vitamins and minerals d. Fats

28. Recommended storage temperature for mango is ....° C
    a. 0 b. 2
    c. 15 d. 8

29. Vitamin C is called as ....
    a. Acetic acid b. Citric acid
    c. Ascorbic acid d. Benzoic acid

30. Banana is best stored at ....° C
    a. 1-2 b. 4-6
    c. 12-13 d. 16-18

31. ° Brix is measured by ....
    a. Salometer b. Hydrometer
    c. Refractometer d. Brixometer

32. Jelly contains .... % sugar
    a. 45-50 b. 35-40
    c. 60-65 d. 68-71

33. .... is done primarily to remove diseased and damaged fruits
    a. Grading b. Sorting
    c. Sorting & grading d. Packing

34. Champagne is a ....
    a. Sparkling juice
    b. Sparkling wine
    c. Sparkling water
    d. Sparkling extract

35. .... is a sparkling clear beverage
    a. Squash
    b. Cordial
    c. Appetizer
    d. RTS drink

36. .... is a non-climacteric fruit
    a. Apple
    b. Litchi
    c. Papaya
    d. Mango

37. Non-climacteric fruits are those which
    a. Ripe on the tree only
    b. Which ripe both on and off the tree
    c. Which ripe off the tree only
    d. Which ripe in storage

38. Pre-cooling is done to ....
    a. Remove field heat
    b. Remove vital heat
    c. Remove sensible heat
    d. Stop respiration

39. .... is the richest source of vitamin C
    a. Orange
    b. Aonla
    c. Seabuckthorn
    d. Guava

40. Potatoes and apple when cut turn brown due to ....
    a. Non-enzymatic browning
    b. Enzymatic browning
    c. Maillard browning
    d. Caramelization

41. Potatoes and apple when cut turn brown due to formation of ....
    a. Melanoidins
    b. Anthocyanins
    c. Carotenoids
    d. Ascorbic aicd

42. Cryo preservation is done at .... °C
    a. 0
    b. -196
    c. 196
    d. 19.6

43 .... contains maximum vitamin C
    a. Guava
    b. Aonla
    c. Papaya
    d. Mango

44 .... contains maximum vitamin A

a. Chiku b. Aonla
c. Papaya d. Mango

45 Vinegar is diluted ....

a. Citric acid b. Lactic acid
c. Acetic acid d. Tartaric acid

46 Guava jelly is preserved by ....

a. Pectin b. Sugar
c. Acid d. Benzoate

47 .... contains maximum vitamin C

a. Tomato b. Cabbage
c. Turnip d. Radish

48 Bitter pit is a postharvest disorder caused due to the deficiency of ....

a. Ca b. Mg
c. B d. Zn

49. .... is suitable for raisin making

a. Perlette Grape b. Thompson seedless grape
c. Anab-e-shahi grape d. All

50. Waxing reduces ....

a. Respiration b. Shelf life
c. Evaporation d. Transpiration

51. .... aims at inactivation of enzymes

a. Blanching b. Pasteurization
c. Sterilization d. Freezing

52. There is rise in respiration rate and ethylene production coincident with ripening in ....

a. Non-climacteric fruits b. Climacteric fruits
c. Temperate fruit d. Tropical fruits

53. .... gas is released during ripening

a. Oxygen b. Ethylene
c. Methane d. Nitrogen

54. Which type of fruits are most suitable for jelly making
    a. Ripe
    b. Unripe
    c. Mature
    d. Mixture ripe and under ripe

55. Sulphur containing foods are processed in .... cans
    a. R enamel
    b. C enamel
    c. Tin cans
    d. Any of the three

56. .... is suitable for gherkin making
    a. Cabbage
    b. Cauliflower
    c. Cucumber
    d. Carrot

57. .... is suitable for cider making
    a. Apple
    b. Pear
    c. Citrus
    d. Papaya

58. .... is suitable for ale making
    a. Ginger
    b. Cauliflower
    c. Cucumber
    d. Apple

59. .... avoids "scurvey disease"
    a. Lemon
    b. Lime
    c. Aonla
    d. All

60. Grading and marketing rules were initiated for ....
    a. Mango
    b. Apple
    c. Grape
    d. Orange

61. Garlic produces odour due to ....
    a. Capsaicin
    b. Allicin
    c. Garlicin
    d. Carotein

62. .... is synonym to holding solution
    a. Vase solution
    b. Pulsing solution
    c. Sucrose solution
    d. Nutrient solution

63. .... is problem in jelly preparation
    a. Weeping
    b. Caremelization
    c. Bitterness
    d. Browning

64. .... causes enzymatic browning
   a. GPO b. FPO
   c. CPO d. PPO

65. Bitterness in bittergourd is due to ....
   a. Momordicin b. Capsicin
   c. Cucurbitacin d. Limonin

66. .... is called as cold sterilization
   a. Freezing b. Irradiation
   c. Chilling d. Carbonation

67. .... is called as lyophilzation
   a. Freezing b. Freeze concentration
   c. Freeze drying d. Frozen storage

68. .... is called as caramelization
   a. Heating sugar with proteins b. Heating sugars without proteins
   c. Heating proteins d. Heating proteins without sugars

69. Fatty foods are spoilt due to ....
   a. Rancidity b. Puterification
   c. Fermentation d. Carbonation

70. Proteins are spoilt due to ....
   a. Rancidity b. Puterification
   c. Fermentation d. Carbonation

71. Carbohydrates are spoilt due to ....
   a. Rancidity b. Puterification
   c. Fermentation d. Carbonation

72. .... is a common refrigerant
   a. Ice b. Ammonia
   c. Ethylene d. CO_2

73. .... has a high respiration rate
   a. Apple b. Walnut
   c. Avocado d. Orange

74. .... is used as preservative in mango products
   a. Sodium benzoate   b. KMS
   c. Sorbic acid   d. Nitrates

75. ... is used as preservative in plum and tomato products
   a. Sodium benzoate   b. KMS
   c. Sorbic acid   d. Nitrates

76. .... is used as preservative in rhododendron products
   a. Sodium benzoate   b. KMS
   c. Sorbic acid   d. Nitrates

77 .... is used as preservative in meat products
   a. Sodium benzoate   b. KMS
   c. Sorbic acid   d. Nitrates

78. .... is used as preservative in bakery products
   a. Sodium benzoate   b. KMS
   c. Sorbic acid   d. Nitrates

79. pH .... is dividing line between acid and non acid foods
   a. 0   b. 7
   c. 4.5   d. 10

80. .... is distillate of wine
   a. Whiskey   b. Rum
   c. Brandy   d. Beer

81. .... participates in alcoholic fermentation
   a. Bacteria   b. Yeast
   c. Moulds   d. Fingi

82. .... is used in alcoholic fermentation
   a. *Saccharomyces cerevisiae*   b. *Acetobacter aceti*
   c. *Bacillus sp.*   d. *Clostridium botulinum*

83. .... is involved in food poisoning
   a. *Saccharomyces cerevisiae*   b. *Acetobacter aceti*
   c. *Bacillus sp.*   d. *Clostridium botulinim*

84. .... is called as vinegar bacteria
   a. *Saccharomyces cereviseae*
   b. *Acetobacter aceti*
   c. *Bacillus sp.*
   d. *Clostridium botulinim*

85. .... is not a fermented product
   a. Cider
   b. Vinegar
   c. Sauerkraut.
   d. Cordial

86. Pectin strength is measured by ....
   a. Salometer
   b. Refractometer
   c. Penetrometer.
   d. Jel meter

87. Fruit pressure / firmness is measured by ....
   a. Salometer
   b. Refractometer
   c. Penetrometer.
   d. Jel meter

88. Sugar strength is measured by ....
   a. Salometer
   b. Refractometer
   c. Penetrometer.
   d. Jel meter

89. Salt strength is measured by ....
   a. Salometer
   b. Refractometer
   c. Penetrometer
   d. Jel meter

90. LAB (Lactic acid bacteria) is used in preparation of ....
   a. Pickles
   b. Vinegar
   c. Cider
   d. Marmalade

91. Acetic acid bacteria is used in preparation of ....
   a. Pickles
   b. Vinegar
   c. Cider.
   d. Marmalade

92. .... is called Father of Canning
   a. Nicholas Appert
   b. Louis Pasteur
   c. Peter Durand
   d. Thomas Saddington

93. Appertization is equivalent term to ....
   a. Pasteurization
   b. Blanching
   c. Sterilization.
   d. Canning

94. ……………….. is suitable for cordial making
a. Apple b. Lime
c. Avocado d. Guava

95. ……………….. is suitable for jelly making
a. Mango b. Guava
c. Apricot d. Litchi

96. ……………….. is suitable for marmalade making
a. Apple b. Mango
c. Citrus d. Banana

97. ……………….. is suitable for puree making
a. Apple b. Mango
c. Citrus d. Tomato

98. ……………….. is suitable for sauerkraut making
a. Apple b. Cauliflower
c. Cabbage d. Banana

99. Which of the following is a component of vase solutions ………………..
a. Glucose b. Fructose
c. Sucrose d. Galactose

100. Cane sugar is predominantly ………………..
a. Glucose b. Fructose
c. Sucrose d. Galactose

101. Which of the following is a component of pulsing solutions ………………..
a. Glucose b. Fructose
c. Sucrose d. Galactose

102 Most suitable temperature for dehydration is ………^{o} C
a. 30-40 b. 40-50
c. 50-60 d. 60-70

103. Which of the following is a biocide in vase solutions
a. Water b. Fructose
c. Sucrose d. 8 HQC

104. .... is called anti sterility vitamin

a. A b. E
c. K d. B

105. .... is called vitamin P

a. Ascorbic acid b. Phenols
c. Flavonoids d. Ttannins

106. .... is precursor of vitamin A

a. Protein b. Catotene
c. Limonin d. Glycine

107. .... vitamins is/ are water soluble

a. A b. B
c. C d. B & C

108. .... vitamins is/ are fat soluble

a. A b. B
c. C d. B & C

109. .... vitamins is/ are fat soluble

a. A b. E
c. C d. A & E

110. Jam can be prepared form pulps of ....

a. Single fruit b. Two fruits
c. More than two fruits d. Any one of the three

111. Colour pigments in plum are due to ....

a. Anthocynin b. Crotene
c. Xanthophyll d. Dye

112. Jelly contains .... % pectin

a. 1 b. 65
c. 33 d. 10

113. .... is a beverage containing 25 % fruit part and 30 % TSS

a. RTS b. Nectar
c. Cordial d. Squash

114. .... is a beverage containing 25 % fruit part and 40 % TSS
   a. RTS b. Nectar
   c. Cordial d. Squash

115. .... is measured in °Brix
   a. Temperature b. TSS
   c. Pressure d. Firmness

116. .... is measured in lbs
   a. Temperature b. Firmness
   c. TSS d. Salt

117. .... preservative is most suited to coloured products
   a. Sorbic acid b. KMS
   c. Benzoic acid d. Nitrates

118. .... preservative bleaches colour
   a. Sorbic acid b. KMS
   c. Benzoic acid d. Nitrates

119. .... is non-climacteric
   a. Litchi b. Plum
   c. Apple d. Banana

120. .... is non-climacteric
   a. Litchi b. Grape
   c. Lime d. All of these

121. .... is not an added preservative
   a. KMS b. Benzoic acid
   c. Citric acid d. Sorbic acid

122. Enzymatic browning occurs when .... is cut and kept open in air
   a. Potato b. Apple
   c. Pear d. All of these

123. Enzymatic browning occurs when .... is cut and kept open in air
a. Cabbage
b. Okra
c. Potato
d. None

124. Browning involves reaction of ....
a. Sugar and vitamins
b. Sugar and amino acids
c. Sugar and minerals
d. Vitamins and minerals

125. .... removes air from the can contents
a. Seaming
b. Exhausting
c. Brining
d. Appertizing

126. Banana when kept in refrigerator experiences ....
a. Freezing injury
b. Chilling injury
c. Heat injury
d. Physical injury

127. .... °C is the suitable temperature for banana storage
a. 0-3
b. 5-8
c. 10- 13
d. -1 – 3

128. .... requires 65% RH for storage
a. Garlic
b. Onion
c. Both
d. None

129. .... is not suitable for storage in ZECC
a. Onion
b. Garlic
c. Potato
d. All

130. Red / blue colour of fruits and vegetables is ....
a. Water soluble
b. Water insoluble
c. Fat soluble
d. All

131. Red / blue colour of fruits and vegetables is due to ....
a. Carotenoids
b. Anthocyanins
c. Chlorophyll
d. Xanthophyll

132. Pungency in chilli is due to ....
a. Chlorophyll
b. Capsaicin
c. Capsoxanthin
d. Xanthophyll

133. .... is not a floral preservative
a. 8 HQC
b. Ag NO_3
c. Aluminium sulphate
d. Sucrose

134. Button hole injury is common during harvesting of ....
a. Mango
b. Malta orange
c. Apple
d. Plum

135. .... is the maximum permissible limit for adding colouring agents
a. 100 ppm
b. 200 ppm
c. 300 ppm
d. 500 ppm

136. The most common method of preservation for grapes is ....
a. Canning
b. Dehydration
c. Pickling
d. Beverages

137. Jelly has a pH ....
a. 2.2
b. 3.2
c. 4.2
d. 5.2

138. .... is a pre-treatment in drying
a. Sulphuring
b. Sealing
c. Clinching
d. Exhausting

139. .... is a pre treatment in canning
a. Exhausting
b. Sealing
c. Clinching
d. All the three

140. Freezing does ....
a. Reduce the No. of m/o
b. Kill m/o
c. Hinder activity of m/o
d. Squeeze m/o

141. .... is also called lyophilization
a. Freezing
b. Freeze concentration
c. Freeze drying
d. Dehydration

142. Jelly contains .... % sugar
    a. 1 b. 65
    c. 33 d. 10

143. Jelly contains .... % water
    a. 1 b. 65
    c. 33 d. 10

144. .... is a beverage containing 10 % fruit part and 10 % TSS
    a. RTS b. Nectar
    c. Cordial d. Squash

145. .... is a beverage containing 20 % fruit part and 15 % TSS
    a. RTS b. Nectar
    c. Cordial d. Squash

146. .... is also called cold sterilziation
    a. Freezing b. Freeze concentration
    c. Irradiation d. Dehydration

147. CIPHET is located at ....
    a. New Delhi b. Ludhiana
    c. Bangalore d. Chandigarh

148. .... is stored in pits
    a. Apple b. Ginger
    c. Orange d. Cabbage

149. Temperate fruits generally do not experience ..... injury
    a. Freezing b. Chilling
    c. Heat d. Physical

150. Pungency in onion is due to ....
    a. Capsacin b. Allyl propyl disulphide
    c. Pungicin d. Liminin

151. Abrasive peeler is used for peeling ....
    a. Tomato b. Potato
    c. Apple d. Citrus

152. GM foods are known to contain new ....
  a. Vitamin b. Protein
  c. Carbohydrate d. Fat

153. Curing is done in ....
  a. Onion and garlic b. Apple and pear
  c. Mango and orange d. Cucumber and squash

154. Waxing reduces ....
  a. Moisture loss b. Shelf life
  c. Shining d. Microbial load

155. Individual shrink wrapping is ....
  a. MA packing b. CA packing
  c. Hypobaric packing d. ZECC packing

156. FPO was passed in the year ....
  a. 1954 b. 1955
  c. 1956 d. 1960

157. PFA was passed in the year ....
  a. 1954 b. 1955
  c. 1956 d. 1960

158. Vinegar contains ....
  a. 4 % Citric aicd b. 4% lactic acid
  c. 4 % acetic acid d. 4% ascorbic acid

159. Respiration is related to ....
  a. Evaporation b. Transpiration
  c. Oxidation d. Decantation

160. m/o cannot grow at alcohol concentrations beyond .... %
  a. 5 b. 9
  c. 14 d. 20

161. .... has TSS of 15%
  a. RTS b. Nectar
  c. Cordial d. Squash

162. Vodka is prepared form ....
   a. Tomato
   b. Apple
   c. Barley
   d. Potato

163. For every rise in altitude of 500 ft / 152 m boiling point of water reduces by about .....°C.
   a. 1
   b. 2
   c. 5
   d. 10

164. In CA storage .... is controlled
   a. Temperature
   b. Humidity
   c. Gases
   d. All the three

165. Hypobaric is .... storage
   a. Low temperature
   b. Low humidity
   c. Low pressure
   d. Low light intensity

166. Freeze drying involves ....
   a. Evaporation
   b. Sublimation
   c. Transpiration
   d. Respiration

167. Ketchup is prepared form ....
   a. Tomato only
   b. Tomato and pumpkin
   c. Tomato and chilli
   d. Both b and c

168. Sauce is prepared form ....
   a. Tomato only
   b. Tomato and pumpkin
   c. Tomato and chilli
   d. All

169. Water has maximum density at .... °C
   a. 0
   b. 100
   c. 4
   d. 10

170. In lye peeling we use ....
   a. NaCl
   b. NaOH
   c. $NaSO_2$
   d. $NaSO_4$

171. Hot extraction in tomato yields ....
   a. More pulp
   b. More colour
   c. More seeds
   d. More pulp and colour

172. Mead is prepared form ....
   a. Honey b. Apple
   c. Barley d. Potato

173. Irradiation is done by using .... rays
   a. UV b. Infra red
   c. Gamma d. Radio

174. .... is recommended for use in fruits and vegetable products
   a. Nitrates b. Benzoates
   c. Sorbates d. All

175. .... is recommended for use in bakery products
   a. Nitrates b. Benzoates
   c. Sorbates d. All

176. Beer is prepared form ....
   a. Honey b. Apple
   c. Barley d. Potato

177. .... is diluted before serving
   a. Squash b. Cordial
   c. Appetizer d. All

178. ZECC was developed at ....
   a. IIHR b. IARI
   c. CPRI d. ICRISAT

179. ZECC fails to work during .... season
   a. Summer b. Rainy
   c. Winter d. Spring

180. Lime juice based RTS beverage has fruit part of .... %
   a. 5 b. 10
   c. 25 d. 20

181. .... is a class I preservative
   a. KMS b. Nitrates
   c. Sugar d. Benzoate

182. There are no maximum limits for .... preservative
   a. Class II
   b. Class I
   c. Class III
   d. None

183. .... is a class II preservative
   a. KMS
   b. Nitrates
   c. Sugar
   d. A & B

184 .... is also called as ripening hormone
   a. Auxin
   b. Cytokinin
   c. Ethylene
   d. Gibberellin

185 .... is ethylene absorbant
   a. Potassium metabisulphite
   b. Potassium permanganate
   c. Potassium chloride
   d. Potassium hydroxide

186 Yeast is used in preparation of ....
   a. Cider
   b. Concentrate
   c. Cordial
   d. Carbonated beverage

187 Bacteria is used in preparation of ....
   a. Cider
   b. Vinegar
   c. RTS
   d. Carbonated beverage

188 .... is called spiced squash
   a. Cider
   b. Appetizer
   c. Cordial
   d. Vodka

189 Peel shreds are suspended in ....
   a. Marmalade
   b. Jelly
   c. Concentrate
   d. Carbonated beverage

190 .... is associated with agricultural exports
   a. NDDB
   b. NABARD
   c. APEDA
   d. NISCAIR

191 Transgenics are also called .... foods
   a. GM
   b. High value
   c. Altered
   d. Genetic

192. .... is recommended for use in meat products
   a. Nitrates b. Banzoates
   c. Sorbates d. All

193. Referigeration is measured in................
   a. Kg b. Quintals
   c. Tons d. G

194. Calcium carbide is used for ....
   a. Drying b. Preservation
   c. Ripening d. Storing

195. Pulsing is done in storage of ....
   a. Fruits b. Flowers
   c. Spices d. Meat

196. .... are the considerations for storage
   a. Temperature b. Humidity
   c. Atmospheric composition d. All the three

197. ZECC maintains ....
   a. Low temperature, low RH b. High temperature and high RH
   c. Low temperature and high RH d. High temperature and low RH

198. Sugar resistant m/o are called ....
   a. Thermophilic b. Psychrophilic
   c. Osmophilic d. Halophilic

199. Salt resistant m/o are called ....
   a. Thermophilic b. Psychrophilic
   c. Osmophilic d. Halophilic

200. Improperly prepared jam is most likely to be spoilt by .... m/o
   a. Thermophilic b. Psychrophilic
   c. Osmophilic d. Halophilic

KEY TO MULTIPLE CHOICE QUESTIONS

1	2	3	4	5
a	a	a	b	b
6	7	8	9	10
a	a	a	b	c
11	12	13	14	15
b	c	b	d	b
16	17	18	19	20
c	b	c	c	b
21	22	23	24	25
d	c	c	b	a
26	27	28	29	30
c	c	d	c	c
31	32	33	34	35
c	c	b	b	b
36	37	38	39	40
b	a	a	c	b
41	42	43	44	45
a	b	b	d	c
46	47	48	49	50
b	b	a	b	a
51	52	53	54	55
a	b	b	d	b
56	57	58	59	60
c	a	a	d	c
61	62	63	64	65
b	a	a	d	a

66	67	68	69	70
b	c	b	a	b
71	72	73	74	75
c	b	c	b	a
76	77	78	79	80
a	d	c	c	c
81	82	83	84	85
b	a	d	b	d
86	87	88	89	90
d	c	b	a	a
91	92	93	94	95
b	a	c	b	b
96	97	98	99	100
c	d	c	c	c
101	102	103	104	105
c	c	d	b	c
106	107	108	109	110
b	d	a	d	d
111	112	113	114	115
a	a	c	d	b
116	117	118	119	120
b	c	b	a	d
121	122	123	124	125
c	d	c	b	b
126	127	128	129	130
b	c	c	d	a

131	132	133	134	135
b	b	d	b	b
136	137	138	139	140
b	b	a	d	c
141	142	143	144	145
c	b	c	a	b
146	147	148	149	150
c	b	b	b	b
151	152	153	154	155
b	b	a	a	a
156	157	158	159	160
b	a	c	c	c
161	162	163	164	165
b	d	a	d	c
166	167	168	169	170
b	a	d	c	b
171	172	173	174	175
d	a	c	b	c
176	177	178	179	180
c	d	b	b	a
181	182	183	184	185
c	b	d	c	b
186	187	188	189	190
a	b	b	a	c
191	192	193	194	195
a	a	c	c	b
196	197	198	199	200
d	c	c	d	c

2. Fill in the Blanks

1. .... is also called Cold sterilization.
2. Extrusion is basically a size .... process.
3. The killing of microorganisms during irradiation of food is mainly due to formation of .... in the food.
4. The action of Pulsed Light Treatment is both .... and ....
5. Heat generation in microwaves is caused due to .... and ....
6. .... is equivalent to radiation sterilization
7. High pressure upto .... cause microbial damage
8. India is one of the largest food producer countries, next only to ....
9. Botanically fruits are ....
10. Strawberries are .... fruits while pineapple is .... fruit
11. Worldwide postharvest fruit and vegetables losses are as high as ....%
12. Maximum % of the cold storage facilities are used for ....crop
13. .... is developed in beans during storage which reduced its quality
14. Green pea becomes less................. during storage
15. Asepsis, drying and irradiation are methods of ....
16. The fundamental principle of preservation of food by application of heat is known as ....
17. FAO has categorized food processing into .... sectors
18. .... involves basic processing of natural produce i.e. cleaning, grading sorting, washing, dehusking etc.

19. .... involves high levels of modifications to considerably alter the natural produce and to make it ready to eat. i.e. ketchup, RTS beverages, ice creams etc.

20. The level of processing in India is ....

21. .... agri-export zones (AEZs) are to be setup for end to end development of exports in India

22. NABARD stands for ....

23. Ripening is that stage which begins with the last stages of .... and lasts till the beginning of ....

24. Respiration is reverse to ....

25. Respiration rate is measured in ....

26. Ratio of carbon dioxide produced to that of oxygen consumed is known as ....

27. RQ for glucose is ....

28. A sharp increase in respiration (respiratory peak) is shown by the increase in production of CO_2 and decrease in internal O_2 concentration in case of .... fruits

29. .... fruits ripen on the tree only

30. The rate of a chemical reaction almost doubles for every 10^0C rise in temperature was described by ....

31. There is an inverse relationship between storage life and ....

32. .... is one gas which is produced during the ripening of fruits.

33. .... is a precursor of ethylene.

34. The concentration of oxygen at which the anaerobic respiration starts is known as ....

35. During aerobic respiration glucose converts into .... via. EMP pathway and then into Acetyl CoA and then brteaks down into ....via TCA cycle.

36. .... is detaching a commodity from the point of its origin.

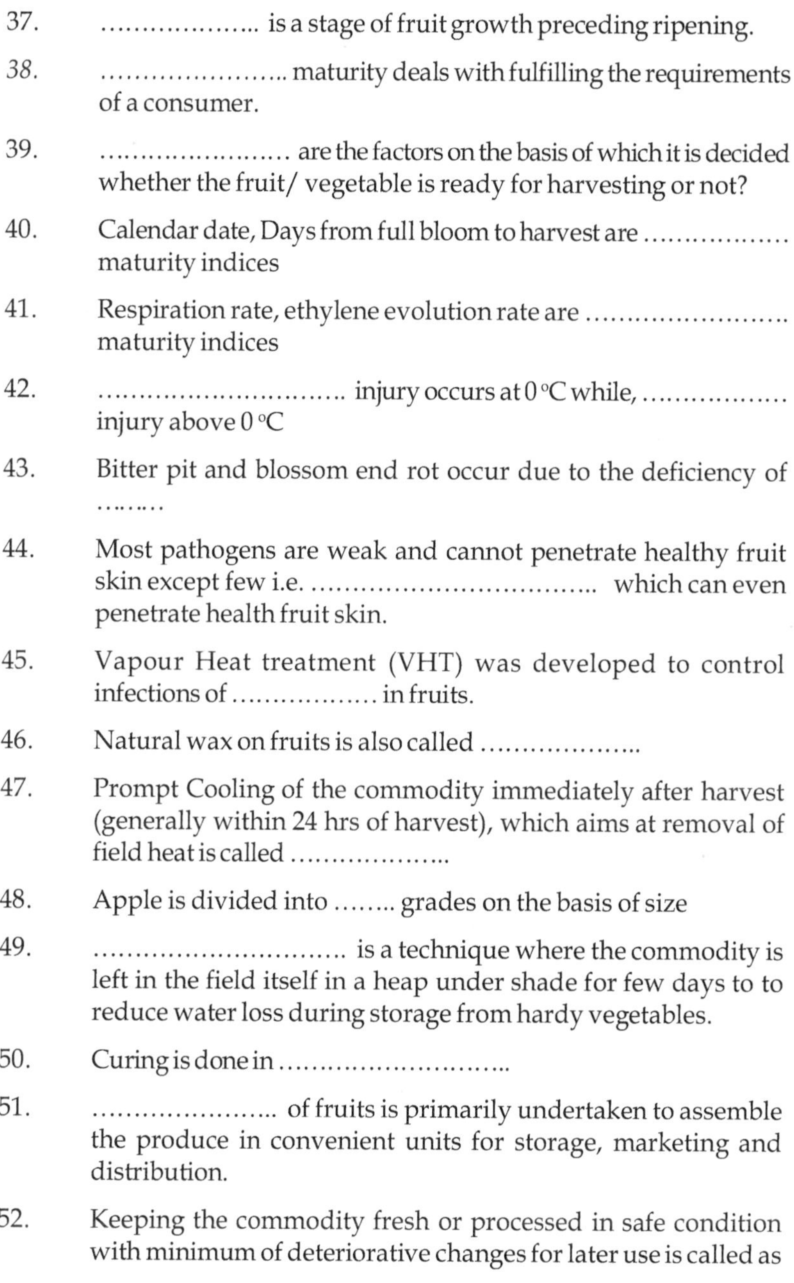

37. .... is a stage of fruit growth preceding ripening.

38. .... maturity deals with fulfilling the requirements of a consumer.

39. .... are the factors on the basis of which it is decided whether the fruit/ vegetable is ready for harvesting or not?

40. Calendar date, Days from full bloom to harvest are .... maturity indices

41. Respiration rate, ethylene evolution rate are .... maturity indices

42. .... injury occurs at 0 °C while, ... injury above 0 °C

43. Bitter pit and blossom end rot occur due to the deficiency of ....

44. Most pathogens are weak and cannot penetrate healthy fruit skin except few i.e. .... which can even penetrate health fruit skin.

45. Vapour Heat treatment (VHT) was developed to control infections of .... in fruits.

46. Natural wax on fruits is also called ....

47. Prompt Cooling of the commodity immediately after harvest (generally within 24 hrs of harvest), which aims at removal of field heat is called ....

48. Apple is divided into .... grades on the basis of size

49. .... is a technique where the commodity is left in the field itself in a heap under shade for few days to to reduce water loss during storage from hardy vegetables.

50. Curing is done in ....

51. .... of fruits is primarily undertaken to assemble the produce in convenient units for storage, marketing and distribution.

52. Keeping the commodity fresh or processed in safe condition with minimum of deteriorative changes for later use is called as ....

53. ...., ... and .... are important storage considerations

54. Generally for most of the fruits and vegetables relative humidity levels of .... are optimum except some commodities i.e. onion, garlic, potato etc. where humidity levels of .... are required.

55. Handling of the produce at optimum (low) temp., RH conditions immediately after harvest till commodity reaches the consumer is called as ....

56. .... are used for storage of ginger

57. .... are underground or partially underground rooms used for storage

58. ZECC works on the principle that "...........................".

59. ZECC is not suitable for............................. (crops)

60. ZECC fails to work efficiently during ....

61. .... is the process of removing heat from an enclosed space, or from a substance.

62. *Cold* is absence of ....

63. Ammonia and tetrafluoraethane are used as ....

64. Tetrafluoroethane is also known as......................

65. Refrigeration load is measured in ....

66. One ***ton of refrigeration*** is defined as the energy removed from .... of water so that it freezes within 24 hours at 0° C (32 °F).

67. One short ton = ... lb

68. 1 tonne of refrigeration is about .... % larger than 1 ton of refrigeration.

69. Energy released by the produce as it respires during storage is called as ....

70. Heat gained by conduction through the building floor, walls, ceiling etc. is called as ....

71. We have better control over the concentration of gasses in ....atmospheric storage than that in .... atmospheric storage

72. Commodity generated MA in sealed packages is also called as.....................

73. .... MA can be generated by pulling out vacuum from the commodity packed and sealed.

74. $KMnO_4$ is used as .... absorber during storage

75. Hydrated lime, activated charcoal, MgO are used as .... absorber during storage

76. Ferrous Oxide is used as .... absorber during storage

77. .... is called low pressure storage

78. .... of fruits is chiefly done to assemble the produce in convenient units for storage, marketing and distribution.

79. Three functions of packaging are ...., ... and ....

80. .... is "to wrap or pack (a commodity) before marketing" or "to package (as food or a manufactured good) before offering for sale to the consumer".

81. The process of packaging of the produce in consumer size units either at producing centre before transport or at terminal markets is called as.....................

82. Individual shrink wrapping of malta, capsicum, and Uncle chips are examples of ....

83. .... (package) comes in direct contact with the food eg can, jar, trays, nets, bottles.

84. Tin cans are made of thin steel plate of low carbon content, lightly coated on either side with tin metal to a thickness of about .... cm.

85. Acid resistant cans are called .... enamel cans while Sulphur resistant cans are called .... enamel cans

86. .... refers to the container that is sealed completely against the entry of gasses and vapours. These are also impervious to bacteria, yeasts, molds and dirt.

87. Hermetic sealing is essential for .... packages.

88. .... are pressurized food packages

89. .... is extending shelf life of fresh or processed food by using various methods and prevention / or reduction of their spoilage.

90. Better the preservation, lesser the ....

91. First principle of preservation deals with prevention or delay of .... decomposition

92. Second principle of preservation deals with prevention or delay of .... decomposition

93. Keeping the microorganisms out is called ....

94. Microorganisms can be removed by ....

95. .... and .... are used to kill m/o

96. .... is responsible for the enzymatic browning of potato or apple

97. Time taken for killing stated number of microorganisms at a certain temperature under defined conditions is called ....

98. Heat treatment at temperatures below 100°C that kills a part but not all of the microorganisms present in the food is called ....

99. Heat treatment generally above 100°C aimed at killing of all the microorganisms is referred to as ....

100. .... is the heat treatment in boiling water at about 100°C given to inactivate enzymes present in the food for about 3-5 min followed by immediate cooling to room temperature.

101. .... is heat preservation of food in hermetically sealed containers.

102. .... refers to freezing in air at -15 to -29°C temperature generally within 3-72 hrs.

103. .... refers to freezing the food at -17.8 to -45.6 °C in less than 30 min.

104. When half of the moisture is first removed by dehydration and then the food is frozen it is called ....

105. Sugar, salt, oil are class .... preservatives

106. Potassium metabisulphite (KMS), Sodium benzoate, Sorbic acid etc. are class .... preservatives

107. .... and .... are used for producing ionizing radiations for use in food irradiation.

108. Pasteurization, filtration, blanching are used in ...... term preservation while sterilization, drying, irradiation are used in .... term preservation.

109. Moisture content in sugar, flour, dry beans is .... therefore are categorized in non perishable foods.

110. Potato, nuts, almond are examples of .... perishable foods

111. pH is the -ve logarithm of .... concentration in a product

112. According to the classification of foods based on pH peas, beans, and corn are grouped in .... foods while pickles, aonla and berries in .... foods.

113. pH .... is regarded as the dividing line between acid and non acid foods, while pH .... is the dividing line between acidic and alkaline

114. During heat processing fruits are heated in boiling water at ....° C while the vegetables (except tomato) at ....° C under pressure because most of the fruits are acidic in nature having low pH.

115. As a thumb rule .... (m/o) does not prefer to grow at low pH while ....do.

116. Low acid foods are first spoilt by .... rather than ....

117. For every rise in altitude of 500 ft or 152 m the boiling point of water reduces by ....°C

118. Pressure at 101° C = ... lb/ in², 115° C = ... lb/ in², 121.1° C = ...lb/ in

119. Tomato contains .... which oxidizes and turns brown when comes in contact with iron.

120. TSS for ketchup and sauce is not less than .... and .... respectively.

121. TSS for tomato juice and soup is not less than .... and .... respectively.

122. Concentrated paste has a TSS not less than ....

123. Acidity for ketchup and sauce is not less than .... and .... respectively.

124. Separation of juice into watery portion and pulp occurs after .... (method of extraction)

125. .... is a concentrated tomato juice or pulp without skin and seeds, containing not less than 25% of tomato solids.

126. .... is prepared only from tomato pulp while .... can be prepared form tomato as well as other pulps i.e. pumpkin, soya, chilli etc.

127. Bag method is a method for adding .... during preparation of sauces and ketchup.

128. Ketchup is preserved by adding .... @ 750 ppm.

129. .... is *a* product prepared by adding spices including chilli, cloves, cardamom, coriander alongwith vinegar and common salt to the tomato juice. Lemon or lime juice etc may also be added in different proportions to suit the palate.

130. Conversion of fermentable carbohydrates into organic acids is the principle of preservation used in ....

131. .... is a pickle prepared from fermented cabbage

132. .... acid fermentation takes place during pickling

133. .... are small sized cucumbers brined for pickling

134. Salt in concentrations above .... % acts as preservative.

135. .... contains 50% TSS and 40% fruit part.

136. Black neck is an important problem encountered during preparation of..................

137. Homofermentative bacteria produces only .... during pickling.

138. Heterofermentative bacteria produces ....,........,...... and .... during pickling.

139. Jam is prepared from fruit.............. while jelly from fruit ....

140. According to FPO Jam should have a TSS not less than .... while jelly not less than ....

141. .... is very important to make gel with acid and sugar

142. When cut .... should retain its shape and show a smooth cut surface

143. .... is a fruit jelly in which the slices of the fruit or the peel are suspended.

144. Mostly marmalades are prepared from ....

145. Sugar under high concentration .... acts as a preservative

146. .... is transparent while .... is not.

147. Finished jam should contain .... % of invert sugar or glucose to avoid crystallization of cane sugar during storage

148. Jam is assumed to be ready when the temperature of the boiling mixture reaches ....° C at sea level

149. At the end point, the weight of the jam is about .... times the weight of the added sugar

150. Normally jams are preserved due to .... and .... activity but .... ppm SO_2 may be added for preservation of jams.

151. Middle lamella of plant cells consist of .... which is insoluble in water.

152. .... when boiled in the presence of acids is hydrolysed to water soluble ....

153. Jelly should have pectin content of .... %, sugar ....% and acid............%.

154. Both ripe and under ripe fruits should be used for preparation of pectin extract as the former yields .... and the latter.....................

155. During alcohol test cooled extract and methylated spirit are mixed in .... ratio.

156. Formation of single transparent lump would recommend addition of extract and sugar in ratio .... while formation of numerous small granule would recommend .... ratio of extract and sugar.

157. Excess of sugar and overcooking may result in .... of jelly

158. Overcooking jelly with excess of acid and low sugar may result in ....

159. .... is the phenomenon of spontaneous exudation of fluid from the gel

160. .... is a jelly like product with shreds of peel suspended.

161. Marmalade is prepared from .... fruits.

162. .... is regarded as a desirable characteristic of marmalades

163. Preserve has TSS not less than .... %.

164. Fruit pieces are kept in .... during preparation and storage of preserves.

165. Process of preservation of fruits and vegetable whole or in pieces, in sugar syrup or brine by processing them in hermetically sealed containers is called as ....

166. .... , a Persian confectioner is known as Father of Canning

167. Canning is also called as.....................

168. .... proved that microorganisms are the real cause of spoilage and by destroying them food can be preserved in suitable containers and introduced the term Pasteurization.

169. Boiling caustic soda (NaOH) @ 1-2 % used for peeling is also called ....

170. .... is the treatment of fruits or vegetables with boiling water or steam for short periods of time followed by immediate cooling prior to canning.

171. .... is the major function of blanching.

172. 100° Salometer reading = ... % salt concentration

173. Maximum solubility of salt in water is .... %

174. The temperature of syrup or brine at the time of can filling should be .... °C.

175. Head space of .... cm should also be kept at the time of filling syrup or brine.

176. Partially seaming the lid to the can by a single first roller operation action of the double seamer is known as ....

177. Clinching is done just before .... during canning.

178. Capacity of A 2½ cans is .... cm^3.

179. Clinching is done by .... roller operation and sealing is done by .... roller operation of the double seamer.

180. Hydrogen swell is a problem associated with ....

181. .... means the removal of water

182. .... is done under controlled conditions while .... make use of renewable energy sources i.e. Sun, Wind etc. for removal of moisture.

183. Microorganisms generally don't prefer to grow at water activity less than ....

184. Reduction of......................... is the principle of preservation in drying and dehydration.

185. Generally dehydration of fruit products is carried out at .... °C temperature.

186. The lesser the RH the .... the drying

187. Heat transfer takes places from one particle to the next particle without the movement of the particles in food is known as .... heating.

188. When heat transfer takes places with the movement of particles themselves it is known as ....

189. Spray driers are used for drying foods i.e. ....

190. Study of inter relationships between temperature and humidity of the air is known as ....

191. Tendency of dried foods to absorb moisture from the surrounding air is called as ....

192. The maximum permissible limit for addition of anti-caking agents in powders is ....% either singly or in combination

193. The process of staking dried fruits in boxes/ bins to equalize the moisture contents before final packaging is called ....

194. Partial removal of moisture by placing fruit pieces in sugar solution followed by final dehydration in air is called ....

195. .... foods contain moderate levels of moisture (20-50%), and a low water activity (unsuitable for the growth of microorganisms) due to high concentration of dissolved solutes.

196. Honey, jam, jelly, cakes, dates, osmo-dried products are examples of ....

197. .... is the measure of effective concentration of water in a substance.

198. Frozen Storage refers to storage at temperatures generally .... °C or below.

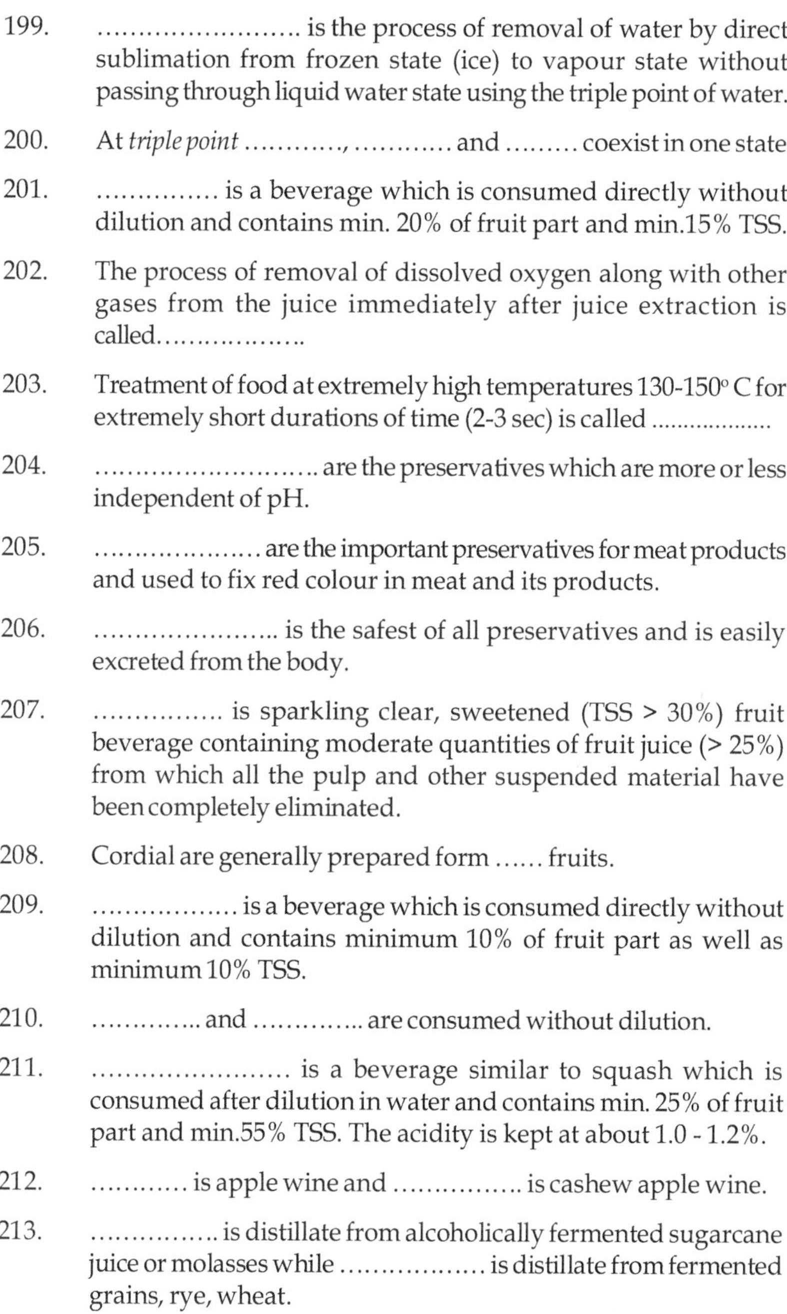

199. .... is the process of removal of water by direct sublimation from frozen state (ice) to vapour state without passing through liquid water state using the triple point of water.

200. At *triple point* ...., ... and .... coexist in one state

201. .... is a beverage which is consumed directly without dilution and contains min. 20% of fruit part and min.15% TSS.

202. The process of removal of dissolved oxygen along with other gases from the juice immediately after juice extraction is called..................

203. Treatment of food at extremely high temperatures 130-150° C for extremely short durations of time (2-3 sec) is called ....

204. .... are the preservatives which are more or less independent of pH.

205. .... are the important preservatives for meat products and used to fix red colour in meat and its products.

206. .... is the safest of all preservatives and is easily excreted from the body.

207. .... is sparkling clear, sweetened (TSS > 30%) fruit beverage containing moderate quantities of fruit juice (> 25%) from which all the pulp and other suspended material have been completely eliminated.

208. Cordial are generally prepared form ...... fruits.

209. .... is a beverage which is consumed directly without dilution and contains minimum 10% of fruit part as well as minimum 10% TSS.

210. .... and .... are consumed without dilution.

211. .... is a beverage similar to squash which is consumed after dilution in water and contains min. 25% of fruit part and min.55% TSS. The acidity is kept at about 1.0 - 1.2%.

212. .... is apple wine and .... is cashew apple wine.

213. .... is distillate from alcoholically fermented sugarcane juice or molasses while .... is distillate from fermented grains, rye, wheat.

214. .... is prepared from barley grains

215. .... is a condiment prepared by alcoholic fermentation of carbohydrates of the starch/ sugar containing commodities i.e. fruits, potato, molasses etc. or followed by acetous fermentation.

216. .... is called vinegar bacteria

217. .... Unit for expressing acetic acid in vinegar.

218. .... is the enzyme which is non- protein in nature

219. Enzymes are very specific to .... and ....

220. .... enzyme is used for breakdown of starch and .... for breakdown of proteins

221. .... enzyme is involved in conversion of sucrose into glucose and fructose.

222. .... is a technique by which enzyme is adsorbed / attached on to a non reactive material and is added with the substrates during the reaction so that it is separable after the completion of the reaction.

223. .... is the deteriorative process which renders the food inedible or results into reduction of quality.

224. .... m/o mostly grow between pH 4-8 and their growth at lower pH becomes difficult.

225. .... are fungal metabolites highly toxic to many animals and potentially toxic to human beings.

226. .... is a mycotoxin produced by *Aspergillus flavus* while .... by *Penicillium sp.*

227. Food poisoning is caused by .... and .... m/o

228. Food infections are caused by ....

229. *Clostridium and Bacillus* are .... forming bacteria

230. Halophiles are .... tolerant and osmophiles are ....loving

231. The word *market* has been derived form Latin word "..............." meaning merchandise, where business is conducted.

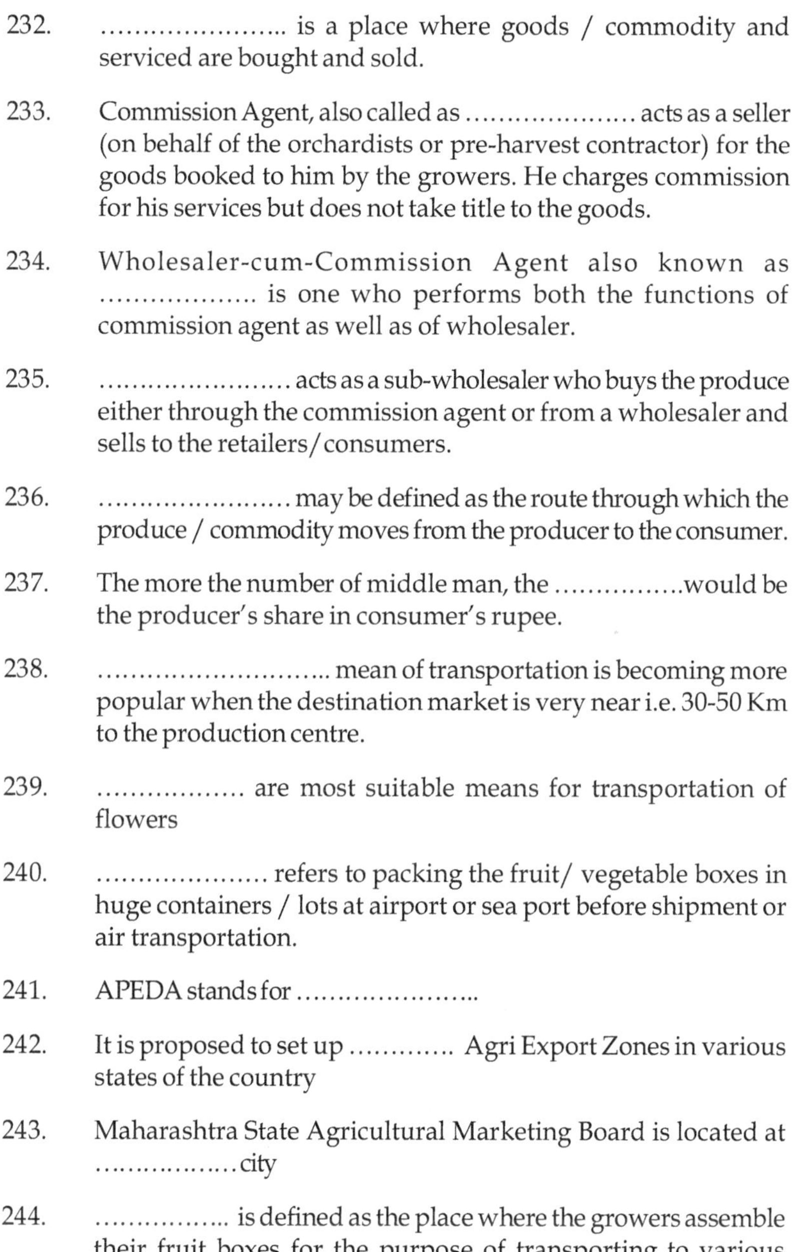

232. .... is a place where goods / commodity and serviced are bought and sold.

233. Commission Agent, also called as .... acts as a seller (on behalf of the orchardists or pre-harvest contractor) for the goods booked to him by the growers. He charges commission for his services but does not take title to the goods.

234. Wholesaler-cum-Commission Agent also known as .... is one who performs both the functions of commission agent as well as of wholesaler.

235. .... acts as a sub-wholesaler who buys the produce either through the commission agent or from a wholesaler and sells to the retailers/consumers.

236. .... may be defined as the route through which the produce / commodity moves from the producer to the consumer.

237. The more the number of middle man, the ....would be the producer's share in consumer's rupee.

238. .... mean of transportation is becoming more popular when the destination market is very near i.e. 30-50 Km to the production centre.

239. .... are most suitable means for transportation of flowers

240. .... refers to packing the fruit/ vegetable boxes in huge containers / lots at airport or sea port before shipment or air transportation.

241. APEDA stands for ....

242. It is proposed to set up .... Agri Export Zones in various states of the country

243. Maharashtra State Agricultural Marketing Board is located at .... city

244. .... is defined as the place where the growers assemble their fruit boxes for the purpose of transporting to various distributing and consuming markets.

245. APEDA, MPEDA and Tea board are .... agencies for export promotion.

246. A slope of .... of an inch per foot is necessary in the processing hall.

247. .... licence must be obtained form Ministry of Food Processing Industries, Govt., of India before starting actual production of the products.

248. LTLT and HTST stands for .... and .... Respectively

249. UHT means ....

250. FPO licence is issued by ....

KEY TO FILL IN THE BLANKS

1. Irradiation
2. Enlargement
3. Free radicals
4. Photothermal, photochemical
5. Rotation of dipoles (dipole polarization) , oscillation of ions (electron polarization)
6. Radappertization
7. 1000 MPa (10000 Bars)
8. China
9. Ripened ovary
10. Aggregate, multiple
11. 30-40 %
12. Potato (58%)
13. Fibre
14. Sweet
15. Preservation
16. Processing
17. Three
18. Primary processing
19. Tertiary processing
20. 1.8-2 %
21. 60
22. National Bank for Agriculture and Rural Development
23. Maturation, senescence
24. Photosynthesis
25. Ml CO_2 $kg^{-1}h^{-1}$(released) or ml O_2 $kg^{-1}h^{-1}$ (consumed)
26. Respiration quotient (RQ)
27. 1.0
28. Climacteric
29. Non-climacteric
30. Van't hoff, a dutch chemist
31. Respiration rate
32. Ethylene
33. Methionine
34. Extinction point
35. Pyruvate, CO_2
36. Harvesting
37. Maturity
38. Horticultural
39. Maturity indices
40. Computational
41. Physiological
42. Freezing, chilling
43. Ca
44. *Colletotrichum*
45. Fruit flies
46. Bloom
47. Pre-cooling
48. Seven
49. Curing
50. Onion, garlic, sweet potato etc.
51. Packaging
52. Storage

53. Temperature, RH and atmospheric composition
54. 85-95%, 65-75%
55. Cool chain concept
56. Pits
57. Cellars
58. Evaporation causes cooling
59. Onion, garlic, ginger, potato etc.
60. July - August (rainy season)
61. Refrigeration
62. Heat
63. Refrigerants
64. Refrigerant 134a
65. Tonnes
66. One ton (2,000 lbs)
67. 2000
68. 10
69. Heat of respiration
70. Conductive heat gain
71. Controlled, modified
72. Passive MA
73. Active
74. Ethylene
75. CO_2
76. Oxygen
77. Hypobaric storage
78. Packaging
79. Containment, unitization and protection
80. Pre-packaging
81. Pre-packaging
82. Modified atmosphere packaging
83. Primary container
84. 0.00025
85. R, C
86. Hermetic
87. Vacuum and gas
88. Aerosols
89. Preservation
90. Spoilage
91. Microbial
92. Self
93. Asepsis
94. Trimming, filtration, centrifugation, sedimentation etc.
95. Heat and irradiation
96. PPO (polyphenol oxidase)
97. Thermal death time
98. Pasteurization
99. Sterilization
100. Blanching
101. Canning
102. Sharp (slow) freezing
103. Quick freezing
104. Dehydrofreezing
105. I
106. II
107. Cobalt - 60, Cesium 137
108. Short, long

109. < 8-10%
110. Semi
111. Hydrogen ion
112. Low acid, high acid
113. 4.5, 7.0
114. 100, 115-121.1
115. Bacteria, molds and fungus
116. Bacteria, molds and fungus
117. 1
118. 1,10,15
119. Lycopene
120. 25%,15%
121. 5%,7%
122. 33%
123. 1%,1.2%
124. Cold pulping
125. Tomato paste
126. Ketchup, sauce
127. Spices
128. Sodium benzoate
129. Tomato cocktail
130. Pickles
131. Sauerkraut
132. Lactic
133. Gherkins
134. 12
135. Chutney
136. Sauces and ketchups
137. Lactic acid
138. Lactic acid, acetic aicd, CO_2, ethanol
139. Pulp, pectin extract
140. 68%, 65%.
141. Pectin
142. Jelly
143. Marmalade
144. Citrus fruits like oranges, lemons etc.
145. >66%
146. Jelly, jam or cordial, squash
147. 30-50
148. 106
149. 1.5
150. High sugar, low water, 40
151. Protopectin
152. Protopectin, pectin
153. 0.5 – 1.0, 60-65, 0.5-1.0
154. Good flavour, good pectin
155. 1:3
156. 1:1, 1:0.5
157. Crystallization
158. Weeping jelly/ syneresis
159. Syneresis
160. Marmalade
161. Citrus
162. Bitterness
163. 68
164. Sugar syrup
165. Canning
166. Nicholas Appert
167. Appertizing

168. Louis Pasteur
169. Lye solution
170. Blanching
171. Inactivation of enzymes
172. 26.5
173. 26.5
174. 79-82
175. 0.32 to 0.47
176. Clinching
177. Exhausting
178. 848
179. First, second
180. Canning
181. Drying/ dehydration
182. Dehydration, drying
183. 0.50-0.65
184. Water activity
185. 50-60
186. Faster
187. Conduction
188. Convection heating
189. Liquids, purees
190. Psychometry
191. Hygroscopicity
192. 2.0
193. Sweating
194. Osmotic dehydration
195. Intermediate moisture foods
196. Intermediate moisture foods
197. Water activity (a_w)
198. -18
199. Freeze drying / lyophilization
200. Ice, water and vapours
201. Nectar
202. Deaeration
203. UHT processing
204. Parabens (para hydroxy benzoic acids)
205. Nitrates
206. Sorbates or sorbic acid
207. Cordial
208. Lime
209. RTS drink
210. RTS drink and Nectar
211. Crush
212. Cider, Fanny
213. Rum, whiskey
214. Beer
215. Vinegar
216. *Acetobacter aceti*
217. Grain strength
218. Ribonuclease
219. pH , substrates
220. α amylase, protease
221. Invertase
222. Immobilization
223. Spoilage.
224. Bacteria
225. Mycotoxins
226. Aflatoxin, Patulin

227. *Clostridium botulinum, Staphylococcus aureus*
228. *Salmonella, Streptococcus, E coil*
229. Spore
230. Salt, sugar
231. Marcatus
232. Market
233. Kacha arhatia
234. Pucca arhatiya
235. Mashakhor
236. Marketing channel
237. Lesser
238. Jeep/ camper
239. Refer vans
240. Palletization
241. Agricultural and processed food products export development authority
242. 60
243. Pune
244. Assembling point
245. Government
246. 1/4th
247. FPO
248. Low temperature long time, high temperature short time
249. Ultra high temperatures
250. Ministry of food processing industries, Govt. of India

References

Anonymous, 1998. Achieve food safety – adopt HACCP. *Bev. Food World,* 25(2): 61-62.

Anonymous, 2005. Report of packaging of fresh fruits and vegetables for exports for Agricultural and Processed Food Products Export Development Authority, New Delhi. Indian Institute of Packaging, Mumbai.

Anonymous, 2005a. Package of Practices of Horticultural Crops. Dr Y. S. Parmar University of Horticulture and Forestry, Nauni, Solan H.P. India p 282.

Anonymous, 2008a. Cold chain. *Agriculture Today Yearbook.* 1: 143-144.

Anonymous, 2008b. Agricultural marketing. *Agriculture Today Yearbook.* 1: 156-157.

Anonymous, 2008c. Food processing. *Agriculture Today Yearbook.* 1: 158.

Cameron, E.J. 1940. Report on canned vegetables. *J.A.O.A.C.* 23:607-608.

Crepaco, A.P.V. 1992. Aseptic processing: ohmic heating. In: Encyclopedia of Food Science and Technology, (Hui, Y.H. eds.), John Wiley & Sons Inc, New York, pp 136-143.

Cundiff, E.W. and Still, R.S. 1972. Basic Marketing. Printice Hall India, New Delhi

Dunn, J., Ott, T. and Clark, W. 1995. Pulsed light treatment of food and packaging. *Food Tech.* 49(9): 95-98.

Dureja, H. Kaushik, D and Kumar, V. 2003. Developments in nutraceuticals. *Indian J. Pharmac.* 35: 363-372.

FAO, 2009. Statistical Database of the Food and Agricultural Organisation, Rome. www.fao.org

Fellows, P. 1988. Food Processing Technology- Principles and Practices. Ellis Horwood Ltd., Chichester, England. p 505.

Fellows, P. 2000. Food Processing Technology - Principles and Practices. 2nd ed. Woodhead Publishing Limited, Cambridge England and CRC Press LLC, USA.

Ferguson, L.R. 2001. Role of plant polyphenols in genomic stability. *Mutat. Res.* 475 (1-2):89-111.

FPO, 1991. The Fruit Products Order 1955. Ministry of Food Processing Industries, Government of India, p 49.

Frazier, W.C. and Westhoff, D.C. 1995. Food Microbiology. 4th edn. Tata McGraw Hill Pub. Co., Ltd., New Delhi. p 539.

Goldberg, I. 1994. Introduction. *In*: Functional Foods, Designer Foods, Pharma Foods, Neutriceuticals. (Goldberg, I. ed.) Chapman & Hall, London pp3-16.

Graham, J. 1984. Planning and Engineering Data, 3, Fish Freezing / FAO Fisheries Circular, No. 771, FAO Rome.

Halsted, C. H. 2003. Dietary supplements and functional foods - 2 sides of a coin. *Am. J. Clin. Nutr.* 77(suppl):1001S–7S.

Harper, J.M. 1979. Food extrusion. *Crit. Rev. Food Sci. Nutr.* Feb:155-215.

Hughes, D. 1982. Notes on ionizing radiation - quantities, units, biological effects and permissible doses. Occupational Hygiene monograph, No.5. Science Reviews Ltd.

Hui, Y.H. 1992. Encyclopedia of Food Science and Technology. John Wiley & Sons Inc, New York.

Hulme, A.C. 1971. The Biochemistry of Fruits and Their Products. Vol 1 & 2, Acad. Press, London.

Joshi, V.K. 1998. Fruit Wines, Directorate of Extn. Edu. Dr Y. S. Parmar Univ. Hort. Fty. Nauni, Solan, HP, India, p 226.

Kader, A.A., Robert, R.F., Mitchell, F.G., Reid, M.S., Sommer, N.F. and Thompson, J.F. 1985. Postharvest Technology of Horticultural Crops. Coop. Ext.Univ. Calif. DANR Berkeley 94720. p192.

Kader, A.A. 1986. Potential applications of ionizing radiation in postharvest handling of fresh fruits and vegetables. *Food Tech.* 40(6): 117-121.

Kumar, K., Dharmandra, D.V. and Dodeja, A.K. 2000. Pulsed electric fields as a means of Food Preservation - An Overview. *Processed Food Industry* (September):15-21.

Kushal, S., Arora, J.S. and Bhattacharjee, S.K. 2001. Postharvest Management of Cut Flowers - A Technical Bulletin No. 10, AICRP on Floriculture, Div., of Floriculture and Landscaping, IARI, New Delhi p39.

Lal, G., Sidappa, G.S. and Tandon, G.L. 1998. Preservation of Fruits and Vegetables. ICAR Pub. New Delhi. p488.

Mercier, C. 1980. Structure and digestibility alterations of cereal starches by twin screw extrusion cooking. *In*: Food Process Engineering (Linko, P., Malkki, Y., Olkku , J. and Larinkari , J. eds.) Vol. I, Applied Science, London pp 795-807.

Mertens, B. and Knorr, D. 1992. Development of non-thermal processes for food preservation. *Food Tech.* 46(5): 124-133.

PFA, 2008. Prevention of Food Adulteration Act, 1954 with Prevention of Food Adulteration Rules 1955. Law Publishers (India) Pvt., Ltd. Allahabad. p534.

Pothakamury, U.R., Barbosa-Canovas, G.V. and Swanson, B.G. 1993. Magnetic field inactivation of microorganisms and generation of biological changes. *Food Tech.* 47(12): 85-93.

Potter, N.N. and Hotchkiss, J.H. 1996. Food Science, 5th edn. CBS Publishers & Distributors, New Delhi, p608.

Rajput, C.B.S. and Babu, R.S.H. 1991. Citriculture. Kalyani Pub. New Delhi p368.

Ranganna, S. 1997. Handbook of Analysis and Quality Control for Fruit and Vegetable Products. 2nd edn. Tata McGraw-Hill Pub. Co. Ltd. New Delhi p1112.

Robinson, D. S. 1986. Irradiation of Foods. *Proc. Inst. Food Sci. Tech.* 19(4):165-168.

Rao, N.M. 2004. Genetically Modified Foods (GMF): An Overview. *Indian Food Industry* 23(2):14-15.

Sadasivam S and Manickam A 2004. Biochemical Methods. 2nd edn., New Age International (P) Ltd. Pub., New Delhi, 256p.

Sahni, C.K., Khurdiya, D.S., Dalal, M.A. and Maini, S.B. 1997. Microwave Processing of Foods-Potentialities and Prospects. *Indian Food Packer* 51(5): 32-42

Senam Raju, M. S. 2002. Fruit Marketing in India. Daya Pub. House, Delhi, p212.

Sharma, S.K., Sharma, P.C. and Kaushal, B.B.L. 2001. Effect of storage temperature and folds of concentration on quality characteristics of Galgal (*Citrus pseudolimon* Tan.) juice concentrates. *J. Food Sci. Tech.* 38(6): 553-556.

Sharma, S.K., Sharma, P.C. and Kaushal, B.B.L. 2004. Storage studies of foam mat dried Hill lemon (*Citrus pseudolimon* Tan.) juice powder *J. Food Sci. Tech.* 41(1): 9-13.

Sharma, S.K., Sharma, P.C. and Kaushal, B.B.L. 2005. Technology refinement for preparation of foam mat dried hill lemon juice powder. *Indian Food Packer* 59(1):75-82.

Sharma, K. 1991. Marketing Management of Horticultural Produce. Deep and Deep Publications, New Delhi, p432.

Srivastava, R.P. and Kumar, S. 2002. Fruit and Vegetable Preservation-Principles and Practices. 3rd edn. International Book Distributing Co. Lucknow, p474.

Swarup, R. and Sikka, B. K. 1987. Production and Marketing of Apples. Mittal Pub. Delhi, p157.

Ting, S.V. and Rouseff, R.L. 1986. Citrus Fruits and Their Products. Marcel Dekker Inc., New York, p293.

Vaidya, D. and Vaidya, M. 2000. Fruit juice and juice beverages. *In*: Postharvest Technology of Fruits and Vegetables -Handling, Processing, Fermentation and Waste Management. (Verma, L.R. and Joshi, V.K. eds.) Indus Pub. Co. New Delhi, pp 681-719.

Verma, L.R. and Joshi, V.K. 2000. Postharvest Technology of Fruits and Vegetables -Handling, Processing, Fermentation and Waste Management. Vol. I & II, Indus Pub. Co. New Delhi, p1222.

Wills, R.B.H., McGlasson, W.B., Graham, D., Lee T.H. and Hall E.G. 1996. Postharvest - An Introduction to the Physiology and Handling of Fruit and Vegetables. CBS Pub. & Distributors, New Delhi, p174.

Zimmermann, U., Pilwat, G. and Riemann F. 1974. Dielectric breakdown on cell membranes. *Biophys. J.* 14: 881-889.

Websites

www.fao.org
www.apeda.org
www.hortibizindia.org
www.indiaagristat.com
www.msamb.com/english/export/globalgap.htm
http://micro.magnet.fsu.edu/index.html.
http://agriculture.house.gov/info/glossary/b.htm
http://www.vuatkerala.org/static/eng/advisory/processing/valueaddition/value_addition.htm
http://www.ibef.org/download/Food_060109.pdf
www.mofpi.nic.in
http://www.mofpi.nic.in/industryspecificinformation/index.htm
http://farzanapanhwar.blogspot.com/2007/08/post-harvest-technology-of-fruits-and.html
http://www.ficciagroindia.com/post-harvest-mgmt/

Annexures

ANNEXURE - I

The Fruit Products Order 1955

The Fruit Products Order 1955 was made by the Central Government in exercise of the powers conferred by section 3 of the Essential Commodities Act 1955.

"Fruit Products" means any of the following articles

i. Synthetic beverages, syrups and sherbets
ii. Vinegar, weather brewed or synthetic
iii. Pickles
iv. Dehydrated fruits and vegetables
v. Squashes, crushes, cordials, barley waters, barreled juice and ready to serve beverage, fruit nectar or any other beverage containing fruit juices or fruit pulp

vi. Jams, jellies, marmalades
vii. Tomato products, ketchup and sauces
viii. Preserves, candied and crystallized fruits and peels
ix. Chutneys
x. Canned and bottled fruits, juices and pulp
xi. Canned and bottled vegetables
xii. Frozen fruits and vegetables
xiii. Sweetened aerated waters with or without fruit juice or fruit pulp
xiv. Fruit Cereal Flakes
xv. Any other unspecified items relating to fruits and vegetables.

Licensing Officer: Director (Fruit & Vegetable Preservation), Food and Nutrition Board, Department of Food, Ministry of Agriculture, Govt. of India, and includes any other officer empowered in his behalf by him with the approval of the Central Government.

Manufacturer: A person engaged in the business of manufacturing fruit products for sale and includes any person who obtained fruit products from another person and packs or labels them for sale. After every two years the Central Government constitutes a committee to be called as *Central Fruit Products Advisory Committee* under the chairmanship of Joint Secretary to the Govt. of India, Department of Food. The Executive Director, Food and Nutrition Board, Department of Food is the Vice Chairman.

THE SECOND SCHEDULE - PART I (A)

Sanitary Requirements of a Factory Manufacturing Fruit Products

The place, where any fruit products are manufactured (hereinafter referred to as the factory), shall comply with the following requirements and in the opinion of the Licensing Officer, shall be fit for manufacturing the item or items for which the licence is granted to the manufacturer.

1. The premises shall be clean, adequately lighted and ventilated and shall be cleaned, if required, by lime washing or colour washing or painting or disinfecting or deodourising.

2. Windows, doors and other openings suited to screening shall be fly proof. The doors should have springs so that they may close automatically.
3. The equipments and the manufacturing premises approved for the manufacture of fruit products shall not be used for the manufacture of other products repugnant to the manufacture of fruit products except under the conditions given as under:

If the licensed premises are used for the manufacture of both fruit products and fish, meat and egg products there shall be a gap of at least one month when the change is made from fish, meat and egg products to the fruit products.

Explanation

The proposed date of changeover from production of the fish, meat and egg products to that of the fruit products, shall be intimated to the concerned Regional Officer in writing and in case of outstation factories, the intimation shall have to be given by registered letter.

4. The premises shall be located in a sanitary place and free from filthy surroundings.
5. All yards, outhouses, stores and all approaches of the premises shall be kept clean and sanitary.
6. The authorized premises shall be so constructed or maintained as to permit hygienic production and all operations in connection with preparing or packing of products shall ba carried out carefully under strict sanitary conditions laid down in the Factories Act 1934, as amended as modified from time to time. The premises shall not be used as or communicate directly with the residential premises
7. Equipment and machinery when employed shall be of such design which will permit easy cleaning. Adequate arrangements for cleaning of containers, tables, working parts or machinery etc. shall be provided.
8. No vessel, container or other equipment, the use of which is likely to ensure metallic contamination injurious to health shall be employed in the preparation, packing or storage of fruit (Copper or Brass vessels shall be always kept tinned. No iron or galvanized iron shall come in contact with fruit products).

9. The water used in the manufacture shall be potable and if required by the Licensing Officer shall be got examined chemically and bacteriologically by any recognized laboratory. The manufacturer will bear the cost of such analysis.
10. There should be efficient drainage system and there shall be adequate provisions for disposal of refuse.
11. Wherever five or more employees of either sex are employed, a sufficient number of latrines for each sex as under shall be provided.

No. of workers	No. of latrines	No. of wash basins
Upto 25	1	1
25 - 49	2	2
50 - 100	3	3
100 and above	5	5

12. Wherever cooking is done in open fire, proper arrangements shall be made for the outlet of smoke and soot.
13. No person suffering from infections or contagious diseases shall be allowed to work in the factory. Arrangements shall be made to get all the workers engaged in the manufacture of fruit products medically examined once in a year to ensure that they are free from infections, contagious and other diseases. A record of these examinations signed by a Registered Medical Practitioner shall be maintained for inspection.

    The worker engaged in the manufacture of fruit products shall be inoculated against the entric group of diseases and vaccinated against small pox once a year and certificate thereof shall be kept for inspection. In case of epidemic all workers should be inoculated or vaccinated.
14. The workers working in processing and preparation shall be provided with proper aprons and headwear which shall be clean. The management shall see that all workers are neat, clean and tidy.

PART I (B)

The factories shall be categorized as under

Category	Installed Capacity per day	Annual Production	Min. area of manufacturing premises excluding store and office space
Large Scale	> 2 metric tonnes	> 250 metric tonnes	300 m²
Small Scale	Upto 2 metric tonnes	> 50 and not exceeding 250 metric tones	
(i) Small Scale Category (A)	Not exceeding 1 tonne	50 tonnes to 100 tonnes	100 m²
(ii) Small Scale Category (B)	Not exceeding 2 tonnes	100 tonnes to 250 tonnes	150 m²
Cottage Scale	-	> 10 tonnes but < 50 tonnes	60 m²
Home Scale	-	Not exceeding 10 metric tonnes	
(i) Home Scale Category (A)		Not exceeding 5 metric tonnes (Primary processing units operating in rural areas for peeling, slicing and bringing of raw mangoes for the manufacture of mango slices in brine for sale to firms licences under FPO)	10 m² (< 5 workers) 20 m² (6-10 workers)
(ii) Home Scale Category (B)		Not exceeding 10 metric tones (Total production of fruit products except canned vegetables)	25 m²

Water

Every licensee shall arrange for at least one kilo litre (1000 Litre) per day of potable water and its availability shall be adequately increased as per production. Free flowing pipe water supply shall be made available in the processing hall.

contd...

PART II

Product	Min TSS %	Min % of Fruit Part	Remarks
Fruit syrup	65	25	
Crush	55	25	
Squash	40	25	
Cordial	30	25	
Unsweetened Juice	Natural	100	
Sweetened Juice	10	85	
RTS beverages including aerated waters containing fruit juice or pulp	10	10 (5% in case of lime juice based RTS)	
Mango nectar	15	20	
Fruit Juice Concentrate	32	100	
Fruit Nectar (excluding orange and pineapple nectars)	15	20	
Orange and pineapple nectars	15	40	
Mango pulp	12		
Canned mango pulp (sweetened)	15		0.3
Flavoured Sweetened Aerated Waters	8		
Sweetened Aerated Water containing fruit juice or pulp or bits	10	10	

PART III

Barley waters (lemon, orange, grapefruit etc.)	30	25	Min. 0.25% barley starch

PART IV

Synthetic Syrups and Sharbets	65		
Ginger cocktail, ginger beer, ginger ale	30		

PART V

Product	Variety	Special characteristics	General characteristics
Bottled or canned fruits	Any fruit of suitable variety	Head space = Not > 1.6 cm Drained weight Not < 50% (40% for berry) Drained weight shall be determined by draining contents for 2 min., on a sieve of dimensions 20.3 x 20.3 cm having 8 mesh /2.5 cm	Only substances that may be added are fruits, sugar, invert sugar, citric acid and water. No preservative shall be added. No artificial colouring matter shall be present, except in case of cherries and strawberries where permitted colour may be added.

contd...

Bottled or canned vegetables	Any vegetable of suitable variety	Head space = Not > 1.6 cm Drained weight Not < 55% (50% for tomato) Drained weight shall be determined by draining contents for 2 min., on a sieve of dimensions 20.3 x 20.3 cm having 8 mesh /2.5 cm	Only substances that may be added are vegetable, sugar, water oil or fat spices, sauce, citric acid and soluble calcium salts. No preservative shall be added. No artificial colouring matter shall be present, except in case of peas where permitted colour may be added.
PART VI and VII			
Jam and fruit cheese	68	45 (25% for raspberry and strawberry)	Only substances that may be added are sugar, dextrose, invert sugar or liquid glucose, flavouring matter, ascorbic acid, citric acid, permitted colours and preservatives.
Fruit jelly and marmalade	65	45	Fruit jellies shall be made from clear fruit extracts. Marmalades shall be prepared from citrus fruit having suspended slices of peel in the finished product. Only substances that may be added are sugar, dextrose, invert sugar or liquid glucose, ascorbic acid, citric acid, permitted colours and preservatives. Should retain flavor and aroma of the original fruit.
PART VIII and IX			
Candied Crystalized or glazed fruit and peel	70 (total sugar) reducing sugars = not less than 25% of total sugars		Only substances that may be added are sugar, dextrose, invert sugar or liquid glucose, citric acid, soluble calcium salts, flavouring matter permitted colours and preservatives.
Preserves	68	55	Only substances that may be added are sugar, dextrose, invert sugar or liquid glucose, flavouring matter, citric acid, ascorbic acid, permitted colours and preservatives.

PART X			
Fruit Chutney	50	40	The acidity and ash contents shall not exceed 2% and 5% respectively. Dry fruit, raisins, spices, salt, onion, garlic, vinegar, acetic acid, colours and preservatives may be added.
PART XI, XII and XIII			
Tomato juice	5		Only substances that may be added in tomato juice are salt not in excess of 5% by weight, sugar, dextrose, malic acid, ascorbic acid, citric acid and permitted colours.. In tomato soup spices, sugar, salt, starch, butter and milk solids may be added.
Tomato Soup	7		
Tomato Puree	9		Only substances that may be added are common salt, citric acid, ascorbic acid, spices, permitted colours and preservatives.
Tomato Paste	25		
Tomato Ketchup	25		Acidity min. 1.0%, Product derived only from tomatoes. Shall not contain any other fruit/ vegetable substance. Only substances that may be added are spices, salt, sugar, vinegar, acetic acid, onion, garlic and preservatives.
Sauce	15		Acidity Min. 1.2%. Product may be from tomato and or any other fruit or vegetable pulp. Only substances that may be added are fruit, vegetable pulps, juice, dried fruits, sugar, jaggery, spices, salt, vinegar, citric acid, acetic acid, malic acid, onion, garlic, flavouring matter, permitted colours other than red or any shade of red colour and preservatives.

PART XIV, XV, XVI and XVII			
Brewed and Synthetic vinegar	Min. 3.75 g acetic acid per 100 ml	Brewed vinegar is liquid derived from alcoholic and acetous fermentation of fruits, malt, molasses, sugarcane juice etc.	Shall contain min. 1.5% w/w of total solids and 0.18 % ash. Shall not contain sulphuric acid, or any other mineral acid, lead or copper, arsenic (exceeding 1.5 ppm) and foreign substance or colouring matter except caramel.
Pickles in vinegar	Min. 2 g per 100 cc as acetic acid in fluid portion		Fluid portion of pickles shall not constitute more than 3/4th of the total content and shall not contain any ingredient other than spices, salt and sugar.
Pickles in citrus juice or brine	Min. 12 % salt. The citric acid contents in pickles in citrus juice shall not be less than 1.2%.		Only substances that may be added are spices, salt, sugar, jaggery, onion, garlic, benzoic acid and soluble calcium salts.
Oil pickle	Prepared from any fruit or vegetable of suitable variety and added any edible vegetable oil like rapeseed, mustard, olive oil etc.		Only substances that may be added are spices, salt, oils, sugar, jaggery, onion, garlic, acetic acid, turmeric, condiments and permitted preservatives.

PART XXII : *Permissible Harmless Food Colours* (The max. limit of any permitted coal tar colours or mixture of permitted coal tar colours which may be added to any fruit products shall not exceed 0.20 g per kilogram i.e. 0.02 %). Only ISI certified colours shall be used.

contd...

PART XXIII

Fruit Product	Preservative	ppm (max.)
1. Fruit and Fruit Pulp or juice	SO_2	1000
2. Fruit Juice Concentrate	SO_2	1500
3. Dried Fruits (apricots, peaches, apples, pears and others) and Vegetables	SO_2	2000
Dried Fruits (raisins or sultanas)	SO_2	750
4. Squashes, crushes, fruit syrups, cordials, barley waters	SO_2 Benzoic Acid	350 600
5. Jam, marmalade, preserve and fruit jelly	SO_2 Benzoic Acid	40 200
6. Sweetened RTS Beverages	SO_2 Benzoic Acid	70 120
7. Pickles and Chutney	SO_2 Benzoic Acid	100 250
8. Tomato and Other Sauces	Benzoic Acid	750
9. Tomato puree and paste	Benzoic Acid	250
10. Syrups and Sharbets	SO_2 Benzoic Acid	350 600
11. Crystallised, glazed, or cured fruit including peel candy	SO_2	150

ANNEXURE - II

Conversion of Units of measurement

Primary Metric Units

Length m
Weight Kg
Volume Litre

Prefixes

Pico = 10^{-12}
Nano = 10^{-9}
Micro = 10^{-6}
Milli = 10^{-3}
Centi = 10^{-2}
Deci = 10^{-1}
Kilo = 10^{3}
Mega = 10^{6}

Length

1 m = 100 cm = 1000 mm = 39.37 inch
1 mm = 1000 μ (micron) = 10,00,000 mμ (milli microns)

1 mμ = 10 Angstrom units (A)
1 inch = 2.54 cm = 25.4 mm
1 foot = 0.305 m = 304.801 mm = 30.48 cm
1 m = 3.279 feet = 1.09361 yards
1 Km = 1000 m = 1093.61 yards
1 yard = 0.9144 m
1 Mile = 1.60934 Km = 1609.34 m
1 Km = 0.62137 miles
8 Km = 5 miles (approx)

Mass

1 grain = 64.799 mg = 0.0648 g
1 ounce (oz) = 28.3495 g = 437.49 grains
1 pound (lb) = 16 oz = 7000 grains = 453.59 g
1 Kg = 10 hectograms = 100 decagrams = 1000g = 2.2046 pounds
1 quintal = 100 Kg
1 metric ton = 0.9842059 ton = 1000 Kg

Volume

1 ml is almost equal to 1cc (cubic centi metre) or 1 ml = 1.000028 cc
1 Litre = 1000 ml = 1000.028 cc (cm^3) = 0.001000028 m^3 = 1000028 mm^3
1 m^3 = 61,023.378 $inch^3$ = 35.3145 ft^3 = 1.3079 yd^3 (cubic yards)
1 m^3 = 999972 ml = 999.972 L = 1000 Litres (approx.)
1 ft^3 = 0.028317 m^3 = 28.317 Litres = 0.028317 Kilo Litres
1 $inch^3$ = 1.6387×10^{-5} m^3 = 16.387 ml = 16387 mm^3

British Conversions

1 Litre = 7.0392 gills = 1.7598 pints = 0.88 quart = 0.22 gallon
1 Gallon = 4 quarts = 8 pints = 32 gills = 4.54596 Litres

US Conversions

1 Litre = 8.4537 gills = 2.1134 pints = 0.1.0572 quart = 0.26418 gallon
1 Gallon = 4 quarts = 8 pints = 32 gills = 3.78533 Litres

Area

1 Ha = 10,000 m^2 = 2.5 Acres (approx.) = 12.5 Bigha (approx.)
1 Bigha = 800 m^2 (approx.)
1 Acre = 4000 m^2 (approx.)
1 m^2 = 10,000 cm^2 = 10,00,000 mm^2 = 10^{-6} Km^2
1 cm^2 = 0.155 in^2 1 in^2 = 6.451613 cm^2

1 Km^2 = 0.38610 sq. mile 1 sq. mile = 2.59 Km^2
1 m^2 = 10.75 sq. feet = 1.196 sq. yds

Temperature

For conversion of temperature recorded on degree Centigrade to degree Fahrenheit

$$^\circ F = \frac{^\circ C \times 9}{5} + 32$$

For conversion of temperature recorded on degree Fahrenheit to degree Centigrade

$$^\circ C = \frac{(^\circ F - 32)}{9} \times 5$$

Freezing Point of water = 0° C = 32° F
Boiling Point of water = 100 ° C = 212° F

Conversion Chart

------- to -------	Multiply by	------- to -------	Multiply by
cm into inch	0.3937	inch into cm	2.54
m into inch	39.37	inch into m	0.0254
m to feet	3.2808	feet into meter	0.3048
Km into mile	0.62137	mile into Km	1.60935
Kg into pounds	2.204621	Pounds into Kg	0.4536
Qunitals into Kg	100	Kg into quintals	0.01
Ounce into g	28.34954	g into ounce	0.03527
Ounce into Kg	0.02835	Kg into ounce	35.27
mm^2 into cm^2	0.01	cm^2 into mm^2	100
cm^2 into m^2	0.0001	m^2 into cm^2	10,000
cm^3 into m^3	0.000001	m^3 into cm^3	10,00,000
ml into l	0.001	l into ml	1000
ml into cm^3	1.000028	cm^3 into ml	0.999972
l into cm^3	1000.028	cm^3 into l	0.000999972
l into m^3	0.001000028	m^3 into l	999.972 (1000 approx.)
% into ppm	10000	ppm into %	0.0001

ANNEXURE - III

Food Additives and Limits of their Use

Food additives	Type of foods	Max. permissible concentrations
Antioxidant : BHA	*Rasogulla* and *vadas*	Not exceeding 0.02 (of the total fat content)
	Whole and partially skimmed milk powder	0.01% (of the finished product)
	Margarine	0.02 %
Anticaking agent: Aluminum silicate	Table salt, onion powder, garlic powder, soup powder	2.0 %
Sweetening agent : Saccharin	Carbonated non-alcoholic drinks	0.01 %
Sequestrant : EDTA	Canned carbonated beverages, salad dressings, margarine and sources	0.003-0.08 %
Colour	Most foods	0.02 %
Flavour : Monosodium glutamate	Meat product, soup powder	0.05 %
Sulphur dioxide or salts of sulphurous acid (Potassium metabisulphite)	Squashes, fruit pulp, crushes, jams, syrups, beer, pickles, beverages etc.	49-3000 ppm
Propionic acid and its salts	Bread and bakery	5000 ppm
Benzoic acid and its salts (Sodium Benzoate)	Chutneys, syrups, squashes, jams, RTS beverages, pickles	50-600 ppm
Sorbic acid	Beverages, cakes and icings, cheese, cider, dried fruits, margarine salad dressings, wine,	200-3000 ppm
Nisin	Cheese	1000 ppm
Nitrites	Meat products	200 ppm
Nitrates	Meat products	500 ppm

ANNEXURE - IV

Richest Sources of Nutrients

a. Richest Common Fruits and Vegetables in Vitamin C

Fruits		Vegetables	
Name	Vitamin C (mg)	Name	Vitamin C (mg)
Seabuckthorn	2000	Parsley	281
Barbados cherry	1540	Coriander leaves	135
Aonla	600	Cabbage	124
Guava	212	Chilli green	111
Cashew apple	180	Amaranth tender	99
Lime	50-63	Brussels sprouts	72
Papaya	57	Knol Khol	85
Strawberry	52		
Lemon	39		

b. Richest Common Fruits and Vegetables in Carotene (precursor of Vitamin A)

Fruits		Vegetables	
Name	**Carotene (µg)**	**Name**	**Carotene (µg)**
Mango	2743	Colocasia leaves	10278
Persimon	2268	Coriander leaves	6918
Apricot	2160	Spinach	5580
Cape gooseberry	1428	Beet (green)	5862
Raspberry	1248	Amaranth tender	5520
Papaya	666	Radish leaves	5295
Loquat	559	Celery leaves	3990
		Mustard leaves	2622
		Fenugreek leaves	2340
		Parsley	1920
		Carrot	1800

c. Richest Common Fruits and Vegetables in Iron

Fruits		Vegetables	
Name	**Iron (mg)**	**Name**	**Iron (mg)**
Currants	8.5	Amaranth tender	25.5
Pistachionut	7.7	Coriander leaves	18.5
Raisin	7.7	Parsley	17.9
Walnut	4.8	Fenugreek	16.5
Almond	4.5	Mustard leaves	16.3
		Brussels sprouts	16.2
		Mint	15.6
		Spinach	10.9
		Colocasia leaves	10.0

d. Richest Common Fruits in Fats and Proteins

Name	Fat %	Name	Protein %
Walnut	64.5	Cashew nut	21.2
Almond	58.9	Almond	20.8
Pistachio nut	53.5	Pistachio nut	19.8
Chilgoza	49.3	Walnut	15.6
Cashew nut	46.9	Chilgoza	13.9
Coconut	41.6	Wood apple	7.1
Avocado	22.8	Coconut	4.5

ANNEXURE - V

Terminology

Aflatoxin – Aflatoxin is a **mycotoxin** produced by two types of moulds i.e. *Aspergillus flavus* and *Aspergillus parasiticus*. 13 different types of aflatoxins are produced in nature with aflatoxin B1 considered as the most toxic.

Antibiotics – Chemical substances produced either by microorganisms or prepared synthetically that inhibit the growth of, or destroy, bacteria are used at therapeutic levels to fight disease in humans and animals.

Biochemical oxygen demand (BOD) – BOD is a measure of the amount of oxygen consumed by natural, biological processes that break down organic matter. High levels of oxygen-demanding wastes in waters deplete dissolved oxygen (DO) thereby endangering aquatic life. This is sometimes called "biological oxygen demand.

Chemical oxygen demand (COD) is a measure of the oxygen consumed when organic matter is broken down chemically rather than biologically.

Codex Alimentarius Commission – It is a joint commission of the FAO (Food and Agriculture Organization) and the WHO (World Health Organization), comprising of 146 member countries. It was created in 1962 to ensure consumer food safety, establish fair practices in food trade and promote the development of international food standards. The commission drafts non-binding standards for food additives, veterinary drugs, pesticide residues and other substances that affect consumer food safety. These standards are published in the "Codex Alimentarius."

Dockage – It is a factor in the grading of grains and oilseeds. Dockage in wheat is would be "weed seeds, weed stems, chaff, straw, or grain other than wheat, underdeveloped, shriveled and small pieces of wheat kernels". These can be removed by sieves and cleaning devices. The term is also used to describe the amount of reduction in price taken because of a deficiency in quality.

Dumping – Practice of selling commodities in a foreign market at a lower price than in the domestic market. Under WTO (World Trade Organization) rules, dumping occurs when the price to the importer is less than the normal price of the product charged to the buyer in the country of its origin.

Effluent – It is a waste, usually liquid, released or discharged into the environment generally from food processing or other industries. It also refers to point source discharges of sewage or contaminated waste waters into surface waters.

Food additives – Any substance or mixture of substances other than the basic foodstuff present added to the food during any phase of production, processing, packaging, storage, transportation or handling.

Food and Drug Administration (FDA) – FDA is a public health agency, responsible for protecting consumers by enforcing the Federal Food, Drug, and Cosmetic Act and several related public health laws. For agriculture, FDA mission is to protect the safety and wholesomeness of food. Scientists test samples to find if any substances, such as pesticide residues, are present in unacceptable amounts. It sets food labeling standards, and ensures that medicated feeds and other drugs given to animals raised for food are not threatening to the consumer's health.

Food-borne illnesses – Illnesses caused by pathogens that enter the human body when carried by food.

General Agreement on Tariffs and Trade (GATT) – It is an agreement originally negotiated in Geneva, Switzerland (1947) to increase international trade by reducing tariffs and non-tariff trade barriers. GATT provides a code of conduct for international commerce and a framework for periodic multilateral negotiations on trade liberalization and expansion. The Uruguay Round of Agreements (1986 to 1993) involving negotiations among over 100 nations established the World Trade Organization (WTO) to replace the GATT institutions officially on January 1, 1995.

Genetic engineering – It refers to the use of recombinant DNA or other molecular gene transfer or exchange techniques to add desirable traits to plants, animals, or other organisms, or to enhance their characteristics or biological processes. Organisms modified by genetic engineering are referred to as transgenic, bioengineered, or genetically modified (GM).

Genetically Modified Organisms (GMO) - Crops carrying new traits that have been inserted through advanced genetic engineering methods (e.g., Flavr Savr tomato, Roundup Ready soybeans, **Bt** cotton, Bt corn). GMO crops are facing resistance from some trading partners, particularly the European Union, that are raising concerns over public health and environmental safety aspects of GMOs.

Generally Recognized as Safe (GRAS) – Food additives that were exempted from the more rigorous regulatory requirements for food additives in 1958. Any substance was accorded GRAS status, if it was generally recognized by experts to be safe under the conditions of its intended use.

Good manufacturing practices (GMPs) – Standards established by the FDA to ensure the quality of marketed products and that products are produced under sanitary conditions. Any regulated product can be designated adulterated if GMPs are not followed properly.

Greenhouse effect – Warming of the earth's atmosphere due to entrapment of infrared radiation emitted from the earth's surface and increased atmospheric carbon dioxide and other gases.

Gross domestic product (GDP) – Measure of the total production and consumption of goods and services in a country. Measures of GDP are prepared on 2 bases, one based on income and one based on expenditures. It is measured on the product side by adding up the labor, capital, and tax costs of producing the output. On the expenditure side, GDP is measured by adding up expenditures by households, businesses, government and net foreign purchases. *Theoretically, these two measures should be equal. However, due to problems collecting data, there is often a discrepancy between the two measures.*

Hazard analysis and critical control point (HACCP) – A production quality control system for minimizing the entry of food borne pathogens, chemicals (toxins, pesticide residues etc.) physical hazards (glass pieces, metal fragments, stones, wood splinters etc.) into the food supply chain in order to protect human health.

High-fructose corn syrup (HFCS) – HFCSs a natural sweetener used as a substitute for higher-cost sugar in soft drink industry are produced by converting a portion of naturally occurring glucose in starch (produced from corn) to fructose. In USA HFCS-55

(55% fructose), which is as sweet as sugar, has almost completely replaced liquid sugar in beverages. HFCS-42 (42% fructose), roughly 90% as sweet as sugar, is mainly used in cereal, baking, dairy and processed foods.

High value products (HVP) – Agricultural products that have high value, (often but not necessarily) due to processing. HVPs are of 3 types (i) semi-processed products, such as fresh and frozen meats, flour, vegetable oils, roasted coffee, refined sugar; (ii) highly processed products that are RTE such as milk, cheese, wine, breakfast cereals and (iii) high-value unprocessed products that are also (RTS/ RTE) i.e. fresh and dried fruits and vegetables, eggs, and nuts.

Irradiation – Process of exposing food to ionizing radiation of various wavelengths in order to destroy contamination from undesirable microorganisms, achieve insect disinfestation or delay maturation.

Long ton – 1 long ton = 2,240 lbs.

Short ton - 1 short ton = 2,000 lbs.

Metric ton = 2,204.62 pounds. Usually abbreviated mt. or MT.

Methyl bromide – A fumigant used to control pests in postharvest storage, of fruits, vegetables, and grain before export.

Most-favored-nation treatment (MFN) – Commitment that a country will extend to another country the lowest tariff rates it applies to any other third country. MFN is a basic principle of the General Agreement on Tariffs and Trade (GATT) (1947).

Mycotoxins – These are toxic substances produced on food by fungi or moulds that may cause sickness in animals or humans. There are about 300 to 400 known mycotoxins, but aflatoxin, vomitoxin, zearalenone, funonisin, T-2 toxin, and T-2-like toxins (trichothecenes) are of main concern.

Nontariff barriers (NTB's) - Any restriction, charge, or policy other than a tariff, that limits access of imported goods. These may include quantitative restrictions, mainly import quotas and embargoes; import licenses; exchange controls; state trading enterprises; bilateral agreements; and certain rules and regulations on health, safety, and sanitation.

Organoleptic – Relating to the senses (taste, color, odour, feel).

Organic foods – Food products produced handled or processed under organic farming, handling and manufacturing processes as defined by several private and state organic certifying agencies.

Parity price – A measurement of the purchasing power of a unit of a particular commodity. Originally, parity was the price per bushel, bale, pound, or hundredweight that would be necessary for a unit of a commodity today to buy the same quantity of other goods (from a standard list) that the commodity could have purchased in the 1910-14 base period.

Sanitary and phytosanitary (SPS) measures and agreements – Measures to protect humans, animals, and plants from diseases, pests, or contaminants. The final act of the Uruguay Round of the Multilateral Trade Negotiations contains "The Agreement on the Application of Sanitary and Phytosanitary Measures." It applies to all sanitary (relating to animals) and phytosanitary (relating to plants) (SPS) measures having direct or indirect impact on international trade. The SPS agreement ensures that SPS measures will not arbitrarily or unjustifiably discriminate against trade of countries other member nor be used to disguise trade restrictions. In this SPS agreement, countries maintain the sovereign right to provide the level of health protection they deem appropriate, but agree that this right will not be misused for protectionist purposes nor result in unnecessary trade barriers. A rule of equivalency rather than equality applies to the use of SPS measures.

Solid waste – It is the non-liquid, insoluble materials including sewage sludge, agricultural refuse, garbage and industrial wastes demolition wastes, and mining residues that may contain complex hazardous substances. Technically, solid waste also refers to liquids and gases in containers.

Technical Barriers to Trade (TBTs) – TBTs a category of nontariff barriers to trade are used to regulate markets, protect their consumers, or preserve their natural resources (among other objectives), but they also can be used (or perceived by foreign countries) to discriminate against imports in order to protect domestic industries. Examples of TBTs, are SPS measures, rules for product weight, size, or packaging; ingredient or identity standards;

mandatory labeling; shelf-life restrictions; and import testing and certification procedures. TBTs have great potential for being misused by importing countries as they are nontransparent (disguised or unclear) obstacles to trade.

Transparency – A WTO principle stipulating that a country's policies and regulations affecting foreign trade should be clearly communicated to its trading partners.

Uruguay Round – The 8th round of multilateral trade negotiations (MTN) conducted within GATT. Launched in Punta del Este, Uruguay, in 1986 and concluded in December 1993, the final Uruguay Round agreement signed in Marrakech in April 1994, embraces 110 participating countries ("contracting parties") and came into effect in 1995. It is being implemented over the period to 2000 (2004 in the case of developing country contracting parties) under the administrative direction of the newly created WTO.

Value-added agriculture – Value added agriculture would be any means to ensure a larger share of farmers in consumer's food dollar or rupee eg. direct marketing; farmer ownership of processing facilities; producing farm products with a higher intrinsic value (such as identity-preserved grains, organic produce, free-range chickens etc.), for which buyers are willing to pay a higher price.

Value-added products – Products that have increased in value in terms of quality and convenience to use because of processing. The terms value-added and high-value are often used synonymously.

Value-based pricing – Instead of simply paying a fixed rate based on the weight of the animals, value-based pricing establishes individual merits of each animal (or lot) purchased, based on quality characteristics such as yield, fat thickness, grade etc to arrive at the price that will be paid. The producer assumes the financial responsibility that the animals, once slaughtered, will meet these criteria.

VAT – Value-added tax.

Source: http://agriculture.house.gov/info/glossary/b.htm

Index

B

C

D

E

F

G

H

I

J

K

L

M

N

O

P

Q

R

S

V

W

X

Y

Z

Zeitfracht Medien GmbH
Ferdinand-Jühlke-Straße 7
99095 Erfurt, Deutschland
produktsicherheit@kolibri360.de